Raising Goats

An Essential Guide on How to Raise Healthy Goats and Tips on Starting a Goat Farming Business from Scratch

Contents

Introduction

This book will help you understand everything you need to know to raise a healthy goat. You will learn all the aspects related to raising a goat successfully at home, understanding the pros and cons, behavior, basic care, housing, diet, and health needs of the animal.

The breeding and selling of animals has happened for many thousands of years, but many modern people are unaware they can raise a farm animal on their land. Lately, though, the methods and manners governing the process have evolved. This book provides all the step-by-step information you will need to raise goats properly.

Like any other domesticated animal, a goat is a precious mammal that needs care and love from humans. Like many other pets, they can share a bond with their owners, but there are also possible pitfalls. You may have a dog or a cat as a pet, but having a goat is very different. Goats can be tricky to handle; many new breeders, companies, or individuals face multiple issues when raising goats for the first time. If you are, by any chance, looking to buy and raise baby goats, then please read this book carefully.

This book will provide you with the common problems, tricks, and caring methods required to raise a baby goat into an adult. It will also give you knowledge and ideas to take care of them as individual creatures and give them the best environment to thrive and grow to be healthy, fit, and active. Whether you want to start up a business, set up a homestead or a farm – or if you just want an unusual pet – this book provides a detailed guide to every aspect of raising, caring, and handling your goats. It also covers all the potential mistakes, common and uncommon, that a new owner may make.

After reading this book, you will have all the knowledge you need before buying a goat and bringing it home. You will also learn various tricks and tips for success; these will be a quick guide when looking for different ways to have fun with your pet. You can expect to learn their basic behaviors, eating habits, shelter requirements, grooming, and training techniques.

In short, the book will help you to be a better owner by learning everything about the animal. This will help you form an everlasting bond with the pet. I am sure you will find many things that surprise and delight as you begin the journey toward understanding these unusual creatures. I'm pleased to have your company as we start our exploration of the remarkable, intriguing—and often downright hilarious—world of goats.

Now, let's get started!

Chapter 1: Why Raise Goats?

Knowing what you would like your goats to help you accomplish is the first step toward determining which breed of goat is right for you. Goats are within the classification, which is known as ruminants. This designation includes those animals that possess four stomachs.

Goats should not be kept alone; they need to be part of a herd. Since they can live longer than seven years, if you are considering getting a herd of goats, be prepared to take care of them over this amount of time.

Goats offer several benefits—food like milk or meat, clothing in fiber or skin, and fuel such as manure.

These animals also offer indirect benefits. One such benefit is how goats have a habit of helping us realize where our food comes from. By being aware of your food source and how to improve it, you can increase your level of self-sufficiency.

Dairy goats can help you reduce or even eliminate your need to purchase dairy. Your goats will need to have kids to provide you with milk. Once this occurs, they will be available for milking for another three years without the need to have further kids.

You must stop milking pregnant goats at the end of the pregnancy during the last two months to help them provide enough nutrition for their kids.

One normal-sized dairy goat can provide three or four quarts of milk per day. Cheese can also be made from the milk. This process usually yields around one pound of cheese for each gallon of milk.

Besides saving money, you will have full control over the quality and safety of the milk and dairy foods you eat by producing your own dairy. And goat's milk is an excellent option for those who may find themselves lactose intolerant. As long as you abide by your local laws, you should be able to sell the excess milk to make a profit or help with the cost of raising your goats.

Goat meat is also rising in popularity. It has always been popular in developing countries due to goats being less expensive than cows. But, as more people move from these countries to more developed countries and bring their customs and recipes with them, goat meat grows in popularity worldwide. Another benefit of goat meat is it is a lean protein source.

When you raise your own goats, you can also control how they are raised. Too often, animals farmed in a confined space may not exercise or eat fresh grass. For peace of mind, raising your own goats for their milk or meat will allow you the benefit of knowing what they consume and any medications they have taken.

Your goats can serve as an alternative source of income, but it is important to do your research before jumping into such an endeavor.

Investigate whether there's a demand for goat meat in your area. This can be done by visiting local auctions or checking your newspaper's agricultural section for current prices.

Next, it is crucial to determine how many animals can be kept on the land you have available. This will be necessary so you can calculate your potential profit.

To raise meat goats, you will first need to go through the purchase process. This means you should check to see how much it costs to procure your initial goats.

There are several ways the slaughtering of your goats can be performed. You can do it yourself, sell them in an auction, have someone come to your farm, or you can take your goats to a slaughterhouse. If you need to take them somewhere, proper transportation must also be considered. Not all slaughterhouses are created equally and identify your nearest USDA-certified facility. Laws and finances will play a big part in your decision-making regarding slaughter methods.

Fiber is another reason to raise goats. Different breeds of goats produce different kinds of fiber, including cashmere, mohair, and cashgora.

Adult angoras can provide anywhere between 8 and 16 pounds of mohair every year, while a young goat, called a kid, can provide around 3 to 5 pounds.

Goats producing cashmere and cashgora cannot provide as much fiber as angoras, but the fiber is worth more. Fiber goats produce fiber, which is used to make things like blankets, clothing, and headwear.

Many people believe there's a breed of goat called cashmere, but this is not the case. Cashmere, instead, refers to the downy hair from certain breeds. Central Asia is the main region where cashmere is harvested. These sweaters are so expensive because you need the fiber from four goats to make one sweater.

Goats can also prove to be handy landscapers. It is common knowledge they are skilled in eliminating weeds. They prefer rough plants. People even rent out their goats to be used by cities to eliminate blackberry bushes and weeds that have taken over a piece of land.

Goats not only eliminate pesky weeds, but they can also reduce the need for herbicides, improve the fertility of the soil, widen the range of plants in the area and reduce the danger of a fire breaking out.

Other animals, like cows or sheep, can coexist with goats in the same area.

Even if you intend to use them for other purposes, they can still be handy when it comes to clearing land. Just keep them safe by using fencing or by having a guard animal.

Goats eat weeds, not grass. Sheep are much more suited to eating grass.

Breeding is usually a big part of owning goats. If you are using them to provide dairy, you will need to breed them to produce enough dairy in the long run. Goats sold or slaughtered will also need to be replaced, which is most easily achieved through breeding.

Money can also be made from selling kids. Another potential income earner is providing a bucking service. Bucking is when you lease out your buck for breeding purposes.

People hire this service for several reasons. Many don't have space or don't want to deal with the complications that result from keeping their own buck. Others are looking for certain genetic aspects that can only be gained through breeding with outside bloodlines.

Besides profit-earning potential, goats make excellent companions.

Miniature goat breeds have grown in popularity, especially among city dwellers. These breeds include pygmy, Nigerian dwarf, and a variety of other miniature dairy breeds. Goats are intelligent creatures that can also serve as comic relief.

Many miniature breeds can be house-broken – and car-broken! By taking them out with you, you are likely to meet new people and can help educate others on the wonderful characteristics of goats.

If you consider keeping one as a pet, it is important to understand they are herd animals. This means you should always have at least two.

Goats can also be leash-trained and can go for walks just as a dog might. This can provide socialization and exercise for both you and your goat.

Once a goat has established a bond with you, it will enjoy spending time with you. They can go hiking with you and help carry your things. They can find food in nature most of the time and are helpful without asking much in return.

Those seriously considering taking their goat packing should select a large, sturdy breed and spend plenty of time training them for the journey.

Goats can even make great therapy animals. They should be specially trained for this kind of work. These animals have been known to help autistic children. Goats can also help improve sensory and social skills in the young and old alike.

Teaching children how to care for and maintain them is a good way for them to learn responsibility. Caring for a goat requires chores twice each day, and children caring for them can be a part of a 4-H project.

Children who participate in a 4-H project involving goat care should expect to create a budget, make a speech, and write reports. They should also demonstrate ordinary care like hoof-trimming, participating in a fair, milking a goat, cheese-making, leash-training, and creating a piece of work for a goat news source.

Before getting a goat, make sure it is the right choice for you. This can be determined by researching them and their habits. And you should spend time with a few goats to get a feel for them and their personalities.

Goats can live upwards of 15 years. If you are buying them as companions or for dairy, be prepared to be responsible for them for that long.

At the minimum, goat owners should spend 30 minutes with them in the morning and again at night. If you have a larger number of goats or are using them for their dairy, meat, or fiber, you will need to spend even more time with them.

They should be fed hay and grass they would typically find in a pasture and should be provided with plenty of water. Time spent with your goats should include cleaning, watering, feeding, and making sure they are safe and healthy. You will probably want to spend as much time as you can with them as you enjoy their personalities.

Those without flexibility in their work schedules still need to handle emergencies, and someone must be available to check on them when you are away.

If you get a goat as a pet, there are a few things you should look out for when selecting your animal—sociability, tameness, and choosing goats without horns. A buck should never be kept as a pet. Instead, get a wether (a buck that has been castrated.) A doe can make a good pet, but wethers require the least maintenance and tend to be sweet.

City dwellers should get miniature goats. Miniatures are easier to handle and are good for those with a physical disability.

Laws and ordinances in your area often regulate keeping goats as pets. Always check these first before buying one. They may be prohibited, and you may receive complaints from neighbors.

Neighborhood dogs can also serve as a threat. If your neighbors are vehemently opposed to having goats in the neighborhood or have an aggressive dog, consider whether owning a goat is a feasible choice. It is always a good idea to familiarize your neighbors with your goats and involve their children in projects with the animals. The more comfortable they feel, the less likely you are to have a problem.

Goat Terminology

Doe - a female goat, also called a doeling when it is young. The term *nanny* is not a proper term for a female goat anymore.

Buck - a male goat, also called a buckling when it is young. *Billy* is no longer an acceptable term for a male goat.

Brood doe - a female goat with desirable genetic traits, kept for breeding.

Kid - a goat less than a year of age.

Yearling - a goat between the ages of one and two.

Wether - a male goat that has been castrated.

Herd - several goats.

Ruminant - an animal with four stomachs and chews cud.

Udder - the organ producing milk. Goats have one.

Teat - what you use on an udder to milk a goat. Goats have two.

Dam - the mother of a goat.

Sire - the father of a goat. This term can also be a verb.

Goats are mammals and, as such, have certain common features with other mammals.

Terms referring to the body parts include—cannon bone, chine, escutcheon, pastern, pinbone, stifle joint, thurl, and withers.

The withers refer to the shoulder area of the spine. The height of a goat is determined by measuring from the withers to the ground.

The thurl refers to the hip joint. Goat aficionados often discuss the levelness of the thurls. The pinbone is the hipbone.

The stifle joint is the knee.

The pastern refers to the flexible portion of the bottom of the leg.

The escutcheon is the area where the udder is on a doe.

The chine is located directly behind the withers on the spine.

The cannon bone refers to the shin.

Because goats are ruminants, they also rely on a plant-based diet. It is important to understand how a goat's digestive system works to make sure your goats are healthy.

Goats are herd animals, so they evolved to live and feed together in a group. This is the environment in which a goat is most happy, and it's important to remember when deciding on how many you should get. At a minimum, you should have two goats. I recommend three, so if something happens to one, you still have the minimum recommendation of two.

While goats are cute, adorable, and very social, they aren't domestic pets, and I don't recommend raising them that way. This doesn't mean you can't love your goats like pets—by all means, do. Even though they can form bonds with humans, they should never be raised alone as a single goat. Humans aren't part of their herd. Though single goats may live happily with other hoofed animals (horses, sheep, etc.), there's no guarantee they will accept other species as part of their herd, so it's important to have two goats.

No matter how affectionate and attentive you are, studies have shown that goats raised without others can become lonely, depressed, and stressed. A stressed or unhappy goat is more susceptible to sickness.

My neighbor used to have a goat as a pet, and she would walk the goat on a leash through the neighborhood. It was adorable until you heard the goat bleating all hours of the night whenever she left the house. The poor thing sounded like it was being tortured. So remember, goats are like cookies - you need more than one!

Which Breed is Right for You?

There are three classifications of domestic goats: dairy, meat, and fiber goats. A few breeds are multipurpose, which means they can be good for both dairy and meat or meat and fiber. Each breed has its area where it shines. Your needs or goals determine the breed you will want to raise, and this section highlights the best in class for each type of goat. We'll also cover other considerations, such as size, cost, and how and where to buy your goats. In the end, you'll be equipped to decide for your homestead or farm business.

Selling Fiber

When we think of fiber animals, the first thing that comes to mind is sheep. Although sheep's wool may be more popular, fiber from goats is considered more luxurious and extravagant. Their fiber is soft, durable, long-lasting, and is used in fine garments.

Selling the fiber is just one of the many ways you can use goats to sustain your self-sufficient life and make money from home. Besides monetizing your herd by selling fiber, you can also save money by making your own clothes. Try your hand at using a spindle and loom or even knitting. You can make beautiful clothes from your goat fiber. Handmade items are also in high demand at artisan stores and markets.

The average cost of wool in the United States is $1.75 per pound, whereas fiber from goats can range from $40 to $190 per pound. Many sheep farmers have raised sheep for meat while raising goats for fiber to make more money.

Fiber Types

There are three different types of fiber you get from goats: mohair, cashmere, and cashgora. Let's take a look at each and discuss the differences between each.

Mohair is the fiber most used in the goat textile industry and comes from Angora goats. It is a strong, beautiful fiber. Turkey was the original producer of mohair, but the demand exceeded its supply, and now the United States is the biggest producer.

Cashmere is the most luxurious textile and highly sought after and, because of its qualities, cashmere also commands higher prices. Many goats, including Pygora and Nigora goats, produce Cashmere.

Cashgora is a happy result of crossbreeding mohair and cashmere goats. Because of this crossbreeding, not every cashgora goat can produce high-quality fiber. Sometimes it's the luck of the draw. Because of this lack of consistency in breed quality, cashgora is considered rare and in demand.

Shearing

When raising fiber goats, you have one of two choices for shearing time: do it yourself or hire someone to do it. Hiring out the task, of course, costs more money. Depending on the type of goat and your location, you can expect to pay an average of $16 to $25 per goat.

But learning how to shear well takes practice. It's not something to be done in a hurry, and you have to have extreme patience. A well-trained shearer can accomplish this task in minutes, while a novice shearer can take one to three hours, which can be stressful on the goat.

Like sheep, Mohair goats need to be shorn twice a year to get their fiber, but cashmere goats must be brushed, making the process more time-consuming.

If you plan on shearing your herd, I suggest hiring a professional for the first time and taking good notes. This way, you can see the process and the techniques they use. Meanwhile, here are a few general tips:

- Expect to harvest 5 to 10 pounds of fiber per adult goat per shearing.

- Angora goats should be shorn in the spring right before kidding season and in the fall before mating season.

- Goats can be sheared when their hair reaches 4 to 6 inches long.

- Have good lighting and your first aid kit handy, including the product Blood Stop.

- Goats need to be cleaned and completely dry before shearing.

- You can use shearing scissors or an electric livestock shear (one designed for sheep).

- Place a drop cloth or tarp down, then your stanchion on top of it to keep your goat in place while shearing.

- Work on one side of your goat at a time.

- Shave as close to the skin as possible without cutting the skin.

- Have two catchment containers available, one for the discarded hair and one for the fiber you need to process. The discarded hair is any coarse or undesirable hair. Use a breathable bag for the fiber for processing.

- Label and tag your bag with the date, age, weight, and name of the goat.

Processing

Once you have fiber from your goats, it has to be made into something for resale. You can sell raw fiber, as many new homesteaders want to do the processing independently or the ready-to-use product. The first step in processing is washing. You need to wash the hair to remove dirt, grease, or impurities. You can wash your goat fiber in the washing machine, but I recommend doing it by hand.

How To Wash Goat Fiber By Hand

Skill Level: Beginner | Estimated Material Cost: $20 | Time: At least 30 minutes

Supplies, Tools, and Steps

- Cloth mesh wash bags

- Washtub

- Hot water

- Detergent or Dawn dish soap

- Drying rack

1. Place the fibers loosely in the wash bag. Do not pack it tight.

2. Fill the tub with water between 145 - 160°F.

3. Add laundry detergent (a quarter cup per pound of fiber) to the water and mix.

4. Gently add the bag of fibers to the water and soak for a couple of minutes.

5. Squeeze out the water.

6. Repeat steps 2 through 5.

7. Once the fiber is clean (dirt-free and bright), you will need to rinse the soap out by following the same processes in steps 2 through 5 (without detergent).

8. Lay the mesh bag on a drying rack to continue draining the water.

9. Once the water is drained, you can remove the fiber and place it on the drying rack until completely dry.

Troubleshooting

Be careful not to wring the fiber or use the agitator in the washing machine. Doing so will cause the fibers to felt (turn felt-like), which is not a good practice.

Spinning

After you wash the fiber, you card or comb it; this straightens the fibers out and removes any debris that the washing may have missed, so the fiber is ready to be spun into yarn.

To spin fiber into yarn, you can use a drop spindle or a spinning wheel. A drop spindle looks like a stick the size of a wooden spoon. It has a spin top on the end and a small hook. It's lightweight, easy to store, and inexpensive, and can be bought for under twenty dollars or so. It is a fun way to learn how to make yarn.

If you plan on processing fiber on a large scale, a spinning wheel might be a better option. Many people sell used spinning wheels online. You can plan on spending a couple of hundred dollars on a spinning wheel.

Selling

As I mentioned earlier, you can sell raw fiber not washed or carded to those wanting to do the processing themselves, large textile corporations, or you can do the processing and sell the completed product. Mohair generally sells for $25 for a four-ounce skein; however, cashgora and cashmere sell for more. Ideas for where to sell your fiber include specialty shops, online marketplaces such as Etsy, local knitting or crocheting clubs, artisan stores, and more.

Additional Services

Here are other strategies for making money with your four-legged livestock. When deciding what kind of business you should set up, it's important to think about your market and how big your products' demand will be. For example, when we lived in the city, we were the only show in town. I could, within reason, create my own market. I would sell my products and farm supplies for top dollar, and my eggs and honey always had a waiting list. To put it simply, it was a matter of supply and demand. Then I moved to the country, where everyone in town raised livestock, and I couldn't give the stuff away. We moved away from our target demographic. So, know where your customers are. There are many places to find customers, including homeschooling groups, health food stores, food co-ops, farmers markets, restaurants that source food locally, other farmers, or online. What is available in your area may inform you of which of these ventures best suits you and your herd.

Selling Your Kids: (goat kids, that is!) Resale value is one reason we decided we wanted a registered herd. With papers, we can sell our registered Nigerian Dwarf does for $200 to $700 each. I would get around $150 for an unregistered goat. Bucks tend to sell for less than does, and wethers go for the least amount since they are not used for dairy or breeding purposes. For the past year, we have focused on growing our herd to increase our milk supply. Now that we have done enough in advance, all the new babies will be sold this year. If you have four does and each doe kids two offspring, you can make anywhere from $1,600 to over $5,000 in one kidding season. Now you can see the math adding up and the potential to make money from selling kids.

Buck Service: If you own a registered buck, you can charge anywhere from $30 to more than $100 per buck service. If your buck comes from show-quality goats, meaning he or his parents won places in goat shows, you can command a higher fee. A one-year-old buck can service ten ladies per season or month. Providing the buck is in good health, a two-year-old buck can service twenty-five females a month, and a three-year-old buck can service up to forty females.

Fertilizer or Compost: Goat pellets or goat berries are great for the garden. They make wonderful fertilizer you can sow directly into the soil. Because of its composition, it doesn't burn your plants, and you don't have to wait a year to apply it. There is hardly any odor, and it doesn't attract flies like cow manure. Gardeners or farmers who work organically are interested in this fertilizer. A farmer friend collected goat manure, then dried it, bagged it into eight-ounce bags, and sold them for $7.95 each plus tax online and at their local nurseries.

Weed Control: All over the country in fire-prone areas, people are singing the praises of goats' ability to clear brush and help prevent the spread of forest fires. Depending on where you live, you may be able to rent goats for weed control and clearing brush. A reasonable price would be a $200 base price, then $125 per week to clear one acre, plus food, travel fees, and other related costs. I've seen others advertise around $1,000 per acre to clear. One important note: because goats are browsers and foragers, not grazers, they do wonders for keeping the brush and weeds at bay but are little good for mowing the lawn.

Goat Therapy: Many animals are used for therapy, helping bring joy, calmness, relaxation, and comfort to humans. Goats are one animal that people use to help them feel better. Goats offer much-needed joy for people who have special needs, seniors, and those with PTSD. Your goats need not be certified to work as therapy animals; they just need to be friendly and enjoy being around people. Another way people are using goats for therapy is by offering goat yoga. All you need are yoga mats, a yoga instructor, and a bunch of bouncing baby goats. People are charging around $45 for a one-hour session to stretch with goats.

Farm Tours and Goat 101: From garden groups to school field trip groups, people of all ages love to visit farms and petting zoos. A goat breeder friend used to offer farm tours and charged $3 per person. She would walk them around the farm, let them pet the goats, observe milking, and even give goat milk samples. Once you have experience under your belt, you can offer classes to other would-be goat owners in goat care, milking, cheese making, goat soap, and much more. Cheese-making classes are $50 to $100 per person, and one-on-one goat mentoring can be offered at $50+ per hour.

Say Cheese: Who doesn't love an adorable spring picture with a baby goat in a field? Spring is the perfect time of year to book photoshoots on your farm with your new baby goat kids. Photographers and individuals alike will love to book photo sessions with your baby goats during the kidding season. Generally, photo sessions start at $50 per hour for photos with baby goats.

Chapter 2: Goat Breeds, Types and Purposes

Breeding goats is an art, but you will need to learn the skills. Knowing what makes one breed distinct from the next helps you deal with a situation should this situation happen.

Miniature Goats

Not all registries recognize miniature goat breeds. We get miniature goats when we crossbreed varieties such as Cashmere, field goats, Angora, Nubian, and others. This goat breed's main traits are that it has a gentle and adorable nature with a high intelligence level. Breeding this variety brings your life a lot of fun and excitement. Miniature goats eat little and require little space. They like to spend their time outdoors and have a good life expectancy of 20 years or more if you care properly for them. They produce a lot of milk for their size, so they are a good investment.

Another significant aspect of miniature goats is how they bond with other farm animals such as cattle and horses. You can allow them to graze together on the field. Goats prefer the company of humans, and so they make good pets. Miniature goats do not develop health problems if there's a change in the weather. As long as you conduct regular health checkups and give them enough food to eat, they will remain healthy. It is possible to buy miniature goats at the tender age of 4-12 weeks. Bottle-feed the kid until it is ready to eat vegetation.

Nigerian Dwarf Goat

The miniature goat breeds do not produce much milk. The Nigerian Dwarf is one such goat with its origins in Africa. They have a very sociable personality, and their coats come in a variety of colors. They produce better-than-average amounts of milk and have a height between 17 and 21 inches. The bucks are 19 to 23 inches in height with a weight of around 75 lbs. We see numerous quadruplets and quintuplets in this breed. They have a maternal attitude and seem to enjoy their kidding experience. The amount of milk from the Nigerian Dwarf is approximately 1 gallon each day. They make a lot of noise and so may not be suitable for city dwellers. Though the meat is edible, people rarely eat them because they are costly and produce little meat.

Pygmy Goat Breed

Another miniature goat is the Pygmy goat. These are stockier and shorter than the Nigerian Dwarf goats. They are not considered dairy goats, but they do produce a fair amount of milk. They have problems kidding due to their stocky nature. The size of the Kinder goat is between that of a Nigerian Dwarf and a Pygmy goat. This goat produces both fiber and dairy. The mini goat is a crossbreed derived from a Nigerian Dwarf buck with a doe from any breed you want to miniaturize. This is useful for those who want to raise goats but do not have enough space for regular goats. Regular de-worming is needed for this breed, along with the trimming of the hooves.

Though the minis will have fewer kids, they will still have triplets and quadruplets. Minis are a great option for people living in the city because of the volume of milk they produce. For an urban lifestyle, the best-suited breeds are Mini Mancha and Oberian because they are the quietest.

Dairy Goats

To get a good dairy goat, use any standard breed. People raise the Pygmy Goat for dairy. For a goat to qualify as a standard breed, the doe must have a minimum weight and height. Their height must be between 28 and 32 inches, but this varies from breed to breed. The bucks must be between 30 and 34 inches in height. All standard goats produce ½ - 1 gallon of milk every ten months. The milk production will be slow initially but build up in speed and then drop back to a low level.

Alpine Goat Breed

Alpine goats are wonderful dairy goats. They are large, and their coats are multicolored. Their ears stand up straight, and because of their big size, they are also used as pack animals. The La Mancha, which is also a dairy animal, has very small ears by contrast, and in a few cases, the ears will appear to be missing. This breed came to the US from Mexico, and we see two types—gopher and elf. These are friendly goats.

Nubian Goat Breed

Nubian is another popular breed of goats. Their ears are floppy, and they increase in numbers through breeding. Nubian goat milk has a good amount of butterfat. They also produce a large amount of milk, so these goats are ideal for making cheese. The Oberhasli is a type of Alpine goat. Their size is medium, and they have upright standing ears. Their coat is reddish and has a black marking. A few will have all black coats. People like this variety because they have sweet temperaments and enjoy being milked.

Other Dairy Oberian Dairy Goats

The colors of the coat of the Sable and Saanen breeds vary. It is not often that you get a dairy goat with a nice coat. Saanen goats have a white coat. These two are the largest of all goat varieties, they give a large amount of milk, and they have a relaxed temperament. They have difficulty keeping their white coats clean, and often, the Saanen goats suffer sunburn. We can recognize the Toggenburg goats by their fawn to chocolate-colored coats. They are the oldest of all dairy breeds. They do not produce a large quantity of milk, and their butterfat content is also average at best.

Meat Goat Breeds

The breeder has as many choices of meat goats as they have with milk goats. Here, the leader of the pack is the Boer. The Boer is sturdy, with its origins in South Africa, and bucks will grow big reaching 300 lbs. in weight and does reach 220 lbs. Boers are usually all white with a brown head but can also be all-brown or all-white. They have long ears and grow fast and gain weight rapidly. Being very fertile, they will produce multiple kids. By nature, they are docile, so it is easy to care for them. This makes them the ideal choice of a goat raised for meat.

Mountain Goats

The Mountain goat is a breed you find in North America. They have a size as big as large wolves and can climb rocky slopes easily. For this reason, they are also called antelope goats. They have a life span of 12-15 years. Because they are big, people rear them for their meat. The adult goat can weigh as much as 220 pounds.

If you want meat goats, the consideration will be different. About 75-80 percent of people in the world eat goat meat. Not all meat goats produce milk. Only the Spanish goat is good for both meat and fiber. We can use meat goats to clean the land. Due to the feral nature of most, they are self-sustaining. We need not spend much time looking after them.

Boer Goat Breed

Boer goats have floppy ears and are large. The bucks weigh between 250 and 360 pounds, while the doe is smaller and weighs between 200 and 250 pounds. Boers are costly and originated from the southern parts of Africa. They are highly adaptable but have many genetic differences, such as abnormal testicles. The Tennessee fainting goat gets its name because it falls down when surprised. Breeders encouraged this defect, so it became popular. The muscles become rigid when the goat falls, so over time, these muscles become strong. That is one reason they make such good meat goats. They are smaller, have a sweet attitude, and weigh between 50 and 70 pounds. Meat goat farmers with limited space will find this goat ideal for him.

Kiko Goat Breed

Kiko is a New Zealand goat, which is a newer breed. It is a meat goat, and it gains weight without feeding. They are even harder than Boer goats. Spanish goats have long horns and are medium-sized. It has a long history, being brought over to the US from Spain during the 16th century. Even though Kiko is a hardy breed, there are fewer breeders, so an effort is needed to preserve them.

Kalahari Red Goat New Moatcashier Red Coat Breed

With origins in South Africa, the Kalahari Red Goat is a meat goat that figures among the top breeds developed like the Savana and Boer goat breeds. Being a new goat breed, it's now quickly gaining recognition in South Africa and other neighboring countries as one of the better meat goats. The first part of the name comes from the Kalahari Desert, while the second part is from its red coat. The Kalahari Desert is spread over vast stretches of South Africa, Namibia, and Botswana.

Many people suspect that the Kalahari Red is a derived breed originating from the Boer goats because of their uncanny resemblance. But blood testing proved that the Kalahari Red is a distinct goat species, not related to any other breeds. The Agricultural Research Council of South Africa conducted this testing. On a comparative note, the Kalahari Red has many advantages over the Boer. We can see this in their camouflage, the kids' survival rates, the goat's hardiness, and the meat's tenderness. Due to its superior meat quality, it is raised extensively in South Africa. This breed is also raised in places like the United States, Brazil, and Australia.

Characteristics of the Kalahari Red

The Kalahari Red is a beautiful animal with a glorious appearance. The breed is large-bodied with a red coat. You might have lighter shades of red and even white but they are not preferred since these colors do not give the goats the needed protection in their native habitat. They have strong herding instincts that serve to protect them. Their ears are long and floppy, and the skin in their neck area is loose. The skin is pigmented, and they have sloping horns that are moderate in size.

They forage throughout the day, even when it is hot. The teats and udders of the doe are properly attached. You can crossbreed any other goat to increase the carcass size and improve the hardiness. The bucks are bigger than the does and have an average body weight of 250 lbs. The doe weighs an average of about 165 lbs.

Use: The biggest use of the Kalahari Red is for meat goat farming. It grows fast and produces very good meat.

Other special properties: the Kalahari Red is a hardy animal that can survive the harsh conditions prevalent in most South African regions. Being very adaptable, they breed well. They are also terrific foragers, feeding on a wide range of plants, grains, and grass. The Kalahari Red will go to vast distances in search of food and water. They require little management because they are resistant to most parasites and diseases. Due to this, one may raise them organically. If you do so, you will get lean meat with excellent texture and taste. They give birth to kids twice every two years, and they do care for the kids carefully.

The kids are born with a strong urge to suck and are strong. The kids grow rapidly, gaining weight at a rate of approximately 3.3 pounds every week. This gives the breeder an excellent profit on his investment.

Other Goat Breeds

Crossover breeding increases the potential of the goats. Examples of these varieties are Savanna, Moneymaker, and Texmaster. Crossbreeding Nubians with Saanens produced the Moneymaker breed in California. Boers were crossbred with Tennessee Fainters to produce the meat goat called Texmaster in Texas.

The Savanna originated in South Africa. It is a new breed that can resist heat well and needs little water. This makes them suitable for breeding in drought-prone regions. These animals are survivors and can adapt to a wide range of oppressive conditions easily.

Fiber Goat Breeds

The fiber goats need more care than the other goats. The size of these goats ranges from small to medium. One type of fiber goat is the Angora, with a coat with long fiber. The fiber is called mohair, which is usually white. Breeders try to make Angora goats in other colors too. One fully-grown Angora will give 8-16 pounds of fiber. This breed is from Turkey, but the United States produces the most Angora fiber. Because they do not have a hardy nature, they must be protected from the elements and extreme cold and heat. They have problems kidding because the does are not naturally maternal.

Cashmere Goat Breed

Cashmere goats are a type of goat rather than a breed. Any goat that produces Cashmere fiber is a Cashmere goat. You can get the starter goats for your Cashmere fiber from Australia. You can use them as meat goats or as fiber goats. Cashmere goats are medium-sized but hardier than Angoras. A buck weighs about 150 pounds, and a doe weighs about 100 pounds. Every December, Cashmere fiber is removed from the goats. Cashmere needs exposure to light for it to grow. The yield is about 4 pounds per year per goat.

Breeding miniature meat goats is not economically viable, but you can find two miniature fiber goats in the US. Crossing a Pygmy with an Angora gives us the Pygora that produces almost as much fleece as an Angora. Many people have them as pets but not for fiber. Even then, it is preferable to shear the fiber twice a year. Then, we have the Nigora breed of goats obtained by crossing a Nigerian Dwarf with a Pygmy. We get both milk and fiber from these goats. But, being a new breed, they are only recently gaining in popularity.

Chapter 3: Housing and Fencing Options

Proper fencing for goats is crucial. Goats are like most animals because they are curious and like to roam. They like their open spaces—the bigger, the better. So, if given a chance to exploit a weakness in their enclosure, you can be sure they'll take it. A strong buck or doe will have little difficulty leaning on the goat fence panels you put up, so much so they will eventually push them down enough to walk over them. It's important to make sure that the ground for your fencing is firm and the materials you use are strong enough for a goat enclosure.

Goat Fencing Basics

Decent options for goat fencing are chain link and woven wire. Remember, if your fence is next to greenery, it will be put to the test, and the goat will push on the fence to get to their food. This type of fence is ideal for young goats and in larger areas.

If using an electric goat fence, the common suggestion is to use one that has seven strands. If it is what you want, you can train your goats to respond well to goat pens with fewer strands. A common modification is to make a typical New Zealand goat fence of only four strands.

Another option is goat pens featuring woven wire-style electric fencing. Most of the feedback is generally positive. There are stories of animals getting tangled in these goat pens and being shocked repeatedly.

To properly utilize electric goat fencing, train the goat by first introducing it to the fence in a smaller surrounded area. This is because if you first introduce the goat to it in a larger area, it will have enough room to charge and get through it as opposed to simply backing away. In a smaller area, the goat will learn to respect the fence because it won't be able to charge through it like it would in a bigger space.

Panel fences are another option and are typically available in three different sizes. The smallest available are Hog panels, which have a height of 3 ft. A pen using this type of panel is best suited for kids since they are not developed enough yet to climb or jump over them. They are also great because the owner can reach in and out of the pen easily.

Another option, and probably the more popular for smaller yards, is cattle panels. These panels are best suited for adolescent and adult goats because of their height (4ft), and they are hard for the animal to jump over. But, when it comes to kids, because of the panel's uniform spacing of 6", some goats may wriggle their way between the panels.

The combo panel is the best panel you can use if you can spend the extra money. It is great because it combines the hog panel's tight spacing at the bottom and the cow panel's hard-to-jump height of 4 ft.

The one catch, and the fence's weakest point, will be the gate clip. You'll want something that you can open easily, one-handed optimally, but, at the same time, you need a clip strong enough to withstand pressure from a goat that wants to escape out of the pen. A lot depends on how many goats you have in the pen, testing its strength. A latch may seem like a good option, but clever goats can open them with ease.

You can try a bungee cord as a quick fix and short-term solution. But they can break down quickly with the pressure goats can put on them and be chewed through with little trouble. Another option and one that is inexpensive and easy to replace is using bailing twine. Remember that you will have a few goats whose sole mission will be to undo the knots with their dexterous little mouths!

Electric Fencing

The following are a set of basic tips for a strong and well-functioning electric fence:

1. Hidden shorts in the electricity can be caused when a staple is hammered through the insulation layer in insulated fences. So, take care when attaching your staples.

2. Bottom wires can ground out if they are too close to snow or grass that is wet, or at the least, their charge can be low as a result. Your best resolution is to be able to shut down these bottom wires.

3. Think twice before running a new fence in tandem with an old one. It's tempting but can cause undesirable shorting out.

4. It goes without saying, but if you're using an electric fence charged by solar panels, make sure they are in a position where they face the direct sun.

5. Use good quality insulators with your fence. Since your fence will face a lot of exposure to the elements, namely the sun, it's important to use an insulator that has been treated to reduce its potential of breaking down.

6. You should make sure to keep at least a 5" gap between wires so they do not cross.

7. Use one type of metal. Electrolysis will corrode your wires. Mixing something like steel and copper wire is likely to do this.

8. Ground your fence properly using many galvanized rods, ideally 6 to 8 ft in length.

9. Fix any damage to the fence as soon as you can. Any sections of wire that get flattened or kinked can wear down and break. Splice your damaged sections by using a square knot tied by hand or a dedicated fence splicer.

10. The spacing of ties and posts is important. If there are too many and they are too tightly spaced, the fence won't stand up to the goats' abuse. There needs to be a little elasticity in the fence.

11. You need to have a fence with a good "kick" to it. A fence using a thin wire that doesn't carry a good charge won't do much to deter your goats much from overcoming it. Even with thicker wires and more expensive insulation, an electric wire fence is still one of the cheaper options to create a perimeter for a big area.

12. Use a voltmeter to check the fence's charge rather than using your hand.

Shelter for Your Goats

The main thing you need is to protect your goats from is the wind and the rain. Otherwise, goats are resilient in terms of both high and low temperatures. If you make sure your goat has shelter from these two elements, you've done your job, and the goat will be happy. This means you can get away with using almost any structure—from a doghouse to a barn, as these will do the job nicely.

A calf hut isn't a bad option because of its low cost, mobility, and ease with which they can be cleaned. There are a couple of cons, though; they deteriorate quickly from the constant exposure to the sun, and they can also be a nightmare to maneuver in when you have to catch your goat.

If possible, a barn with a mix of a concrete floor for human traffic and a slightly lower dirt floor for the goats is probably the best scenario. The dirt floors provide absorbency for waste and make it easier to clean up, while the concrete paths make it a mobile space for owners.

Goats are also great because they don't mind being near each other. But that doesn't mean you need not provide space for them in their living quarters. It is recommended to have about 16 square feet for every fully-grown goat you have. This is to make sure that waste is spread out and will help to maintain cleaner bedding. If adequate space isn't provided, the risk of different diseases like parasites increases dramatically.

A Word About Flies

I think it goes without saying that wherever there's livestock, there *will be flies.* But that doesn't mean you shouldn't do what you can to reduce their numbers. Flies are naturally attracted to the ammonia present in the waste of animals. They lay their eggs and become grown in only a short time; within a single summer season, you're likely to run through many generations of flies.

Your best bet against flies is a mixed approach where you might employ home methods along aided by a pest control expert. The easiest way combines sticky flypaper and flytraps. These will contain a liquid attractant, and the top allows the flies in but not out again, so eventually, they will drown. With the various birds, bats, and insects that are the natural predator of flies, you should hopefully be able to keep them at bay.

Having a good enclosure and shelter for goats is important and something that you should put a little forethought and research into. Go for quality when selecting materials to construct your fence, whether you go electric or not, as any feeble fence's weakness will be exploited by your goat's tenacity. When dealing with shelter, the idea is the same. Do the best with what you've got, but remember that the less expensive options usually come with their inherent flaws.

Chapter 4: Building Your Goat Barn

Our goats need a safe place to call home, a place that protects them from the elements and predators. A goat shelter—or what we like to call a goat manger—can be as elaborate or as simple as you want it to be. You may need to set up a temporary shelter when you first get your goats, but long-term housing plans should begin before you purchase your herd.

Essential Components of a Shelter

Your goat shelter needs to keep your goats dry, give them somewhere to go out of the elements and keep them safe from predators. At the very least, this means goats need a roof and three walls. Recall, too, that no matter what your shelter looks like, each goat in your herd should have at least 20 square feet of living space inside the shelter. Let's take a closer look at the essential components.

Roof: Every goat shelter needs a roof free of leaks.

Walls: The walls should be sturdy enough to keep out wind and rain.

Flooring: The best flooring for a manger is dirt or gravel. Concrete is too cold, and wood absorbs odors. Dirt and gravel allow the manure to disperse and decompose. On top of this, you need a couple of inches of hay, wood shavings, or straw to help provide warmth and absorb smells.

Room and board: If you plan to breed goats or keep both does and bucks, you need separate living quarters. The dividers need not be fancy, but they must be goat-proof, meaning they separate your goats, so they don't have access to each other. You can use wood pallets, wire fencing, wood planks, or tin to divide stalls.

Birthing area: If you plan to breed goats, you'll need a birthing area separate from the main living quarters, away from the rest of the herd. This area provides mama with privacy, safety, and time to bond with her kids. This area should be at least 4 ft by 6 ft, well-insulated with hay, straw, or wood chips, and is free from drafts and the elements. You will also need room for feed and water buckets.

Sleeping and Bedding

Bedding helps keep your goats dry, absorbs ammonia from their manure, and provides warmth. You want something absorbent, soft, and cost-effective. Common bedding materials include:

Hay or straw. This is my preference. It's not as absorbent as the other two, but I always have it on hand to feed the goats and pigs.

Pine shavings. Make sure it's pine and not cedar since cedar can cause issues with goats.

Wood pellets. These are the same pellets used for horse bedding or wood pellet stoves.

Add a couple of inches of bedding to the entire floor, and on any shelf they lie on.

Keeping the Shelter Clean

How many goats you own and the season will determine how often you need to clean out the shelter. We have nine Nigerian Dwarf goats, and we clean out their shelter every other week in the summer to help keep the fly population under control. We use the deep-bedding method in the winter, which means I just keep adding fresh bedding to their existing bedding without removing the old bedding. The decomposition of the old bedding gives off heat, which helps keep the goats warm in the winter months.

Your goat shelter should smell "goaty" or musty. It should never smell like ammonia or burn your eyes. If it does, it is time to clean your goat shelter and replace the old bedding with new bedding.

How to Build a Temporary Shelter

Skill Level: Beginner | Estimated Material Cost: $100 or less | Time: 2 hours

This plan is for a shelter that should be used only for temporary housing until the permanent shelter can be built. You may also need a temporary shelter when moving goats from one area to another on your land or when separating goats. This shelter should have a dirt floor.

Supplies, Tools, and Steps

- 2 (4' × 16') welded wire cattle panels
- Heavy-duty zip ties
- 5 (4') T-post stakes
- 1 (16' × 20') heavy-duty tarp
- Hammer

1. Lay both cattle panels flat on the ground side by side to cover an 8 × 16–ft area. Then push one panel over the other, so they overlap

by 4 in (one square of the panel). Use the zip ties to secure the panels together every three squares.

2. Using two of the T-posts, drive down each post into the ground about 1 ft deep where you want one side of your shelter. Make sure the posts are firm and solid in the ground and on either end of the 8-ft section.

3. Place the shorter end of the cattle panels up against the T-posts secured in the ground. (It's easier if you have a helper.)

4. Hold the opposite end of the cattle panels and slowly walk the end toward the side, pressing against the T-post. This will form the panels into an arch.

5. After the arch is made (arch should be about 9 ft wide), one person continues to hold the arch in place while the helper drives the two remaining T-posts into the ground on the outside of the panel to secure the loose side.

6. Cover the arch with the tarp and secure it to the panels with zip ties. Completely cover the back and sides of the shelter, leaving the front open.

7. Add straw to the floor and place a wood shelf or folding plastic table inside for your goats to lie on.

Troubleshooting

Goats are curious characters, and they like to eat things they shouldn't, including the tarp for their housing. Try to make sure that you secure all loose ends and have no overhang that is not secured, or they will eat it.

How to Build a Permanent Shelter

Skill Level: Beginner | Estimated Material Cost: $500 or less | Time: 2 days

This shelter considers the long-term needs of your goats and will accommodate a bigger herd. You'll start by framing three sides. Next, you'll work on the roof. The 2" × 4" × 10' boards are your roof rafters. The roof will be elevated in the front to help with runoff, and you will have a 1-ft overhang on the front and back. This shelter will be wide enough you can divide it into separate living quarters if needed. If you build this shelter using wood screws instead of nails, it is possible to disassemble the structure and move it if you need to. This shelter should have a dirt floor.

Supplies and Tools for the Framing

- Disposable paintbrush
- Tar
- 3 (4" × 4" × 8') wood posts
- 3 (4" × 4" × 10') wood posts
- Post hole digger
- Level
- Measuring tape
- 2 (2" × 6" × 16') boards
- 1 box (3½") decking screws
- Electric drill
- 4 (6' × 8') wood fence panels

For the Roof

- 10 (2" × 4" × 10') boards
- 6 (1" × 4" × 16') boards
- 8 (26" × 10') tin roofing sections
- 1 bag screws

To Make the Frame

1. Measure and mark off an area for your shelter. Each side will be 8 ft long, and the front and back will each be 16 ft wide. Choose a flat and level place with the opening facing south, if possible.

2. Using the disposable paintbrush, paint the bottom 2 ft of each 8-ft and 10-ft 4" × 4" post with the tar, and then discard the brush. The tar protects the wood post from rotting in the ground.

3. If you are looking at your goat shelter, you will start your building at the back-left corner.

4. Using the posthole digger, dig a hole 2 ft deep. Place one of the 8-ft posts, tar-side down, in the hole.

5. Use the level to check for plumb on the side and front of the post. Once you are plumb, backfill the dirt and pack it firmly around the post. Check for plumb again.

6. From the post's outer corner, measure over to the right 8 ft for the second hole. Repeat steps 4 and 5, placing the second post at 8 ft center (the center of the second post should be 8 ft from the outer corner of the first).

7. From the center of the second post, measure out 8 ft to the right. Repeat steps 4 and 5, using the last 8-ft post.

8. Pull the measurement from the third post's outer edge to the first post's outer edge; the total measurement should be 16 ft.

9. Measure from the outer edge of the first post forward 8 ft. Repeat steps 4 and 5 using a 10-ft post. This will become the corner post for the front of your shelter.

10. Measure from the outer left corner 8 ft. Repeat steps 4 and 5 with the second 10-ft post, placing it at 8 ft center.

11. From the center of the second post, measure out 8 ft to the right. Repeat steps 4 and 5 using the last 10-ft post.

12. You now have six posts in the ground. The three back posts are 6 ft high, and the three front posts are 8 ft high.

13. Position one 2" × 6" against the backside of the 4" × 4" posts, flush with the top of the posts, and secure it with the decking screws; this is the shelter's back cross support.

14. Secure the remaining 2" × 6" board at the top of the front 4" × 4" posts.

15. The top of each 2" × 6" board should be flush with the top of your 4" × 4" posts.

16. Install the fence sections to the back and sides of the 4" × 4" posts using decking screws. You now have all three walls installed in your shelter.

To Make the Roof

17. Starting with the front left, lay the first 2" × 4" on top of the two 16-ft cross-support boards, leaving a 1-ft overhang on both sides of the shelter (front and back). Secure it with decking screws. Make sure to line up the outer edge of your 2" × 4" with the 16-ft cross support's outer edge.

18. From the outside left side of the first rafter, measure over 2 ft and install the second 2" × 4" as you did in step 1. Repeat this process with the remaining 8 (2" × 4") boards until you have ten rafters.

19. Next, install the tin supports, which lie lengthwise across the rafters. Install the first 1" × 4" board flush with the front edge of the front rafter and secure with decking screws.

20. From the front edge of this first tin support, measure backward 2 ft and install the second 1" × 4" board and secure it with decking

screws. Repeat this process with the remaining 4 (1" × 4") boards until you have six tin supports.

21. Secure the tin roofing sections to the supports using tin screws.

Troubleshooting

The roof must be watertight. The tin screws have a rubber gasket on them to help prevent water leaks. You can add silicone caulk to the outside of the screws for an extra level of water protection on the roof.

How to Clean Your Goat Shelter

Skill Level: Beginner | Estimated Material Cost: $10 per month per 8' x 16' shelter | Time: 30 minutes

Supplies, Tools, and Steps

- Dust mask or respirator
- Gloves
- Shovel
- A large trash can or wheelbarrow
- Fresh bedding (pine shavings)

1. Put on your mask and gloves, and make sure all the animals are out of the goat shelter.

2. Shovel out the old bedding, placing it into the trashcan or wheelbarrow for composting.

3. Remove any spider webs and inspect for rodents.

4. Once you've removed all the used bedding, let the shelter air out for about 1 hour.

5. Add a couple of inches of new bedding. Compost the manure and old bedding.

6. Rinse off the shovel and spray with a disinfectant such as bleach.

Chapter 5: Feeding Your Goats

In the wild, goats are browsers rather than grazers. Wild goats eat mostly trees and other food sources that are off the ground. These plants often have deep roots that bring minerals up from the subsoil. We can learn from this and arrange it so our goats can eat mineral-rich food, which is off the ground in a backyard environment. When we learn from nature and apply it to the backyard environment, our goats can enjoy healthy lives free from parasites and diseases.

Goats are highly prone to parasites if they eat their food too close to the ground. For grazing goats, the pasture should be at least 16 inches (15 cm) high if you want them to eat any of it. Hay and other brought-in foods will need to be kept off the ground too, either with a hay feeder or manger or with cheap hay bags or rubber tubs that can attach to the fence. Hay bags can be found wherever you can find horse supplies. Avoid the net style if your goats have horns.

The strict butting order in goat herds and the amount of time that they spend eating means that you must make sure each goat can access food when they want to. Providing a separate hay bag or tub for each adult goat will achieve this, as will a large manger with more than enough space for your entire herd at once.

Goats can be very fussy with hay. Lucerne, clover, and other legume hays are usually easy to find and favored by goats (avoid red clover hay for white goats, it may be too high in copper for them). Second-cut grass hay, carefully made from fertile pastures, can be acceptable. If horses are fond of the hay you're looking at, your goats will probably be as well. They will also favor grain crops that have been baled as hay well before the seed heads are mature while the grass is still soft.

If you're unsure about hay, it's best just to buy one bale and offer your goats a little to see if they like it. Never feed them moldy hay, and try to keep it out of direct sunlight in a well-ventilated area. The traditional hayloft of a barn is the ideal place to store hay as it gets plenty of ventilation. Garages and carports are also fine for storing hay.

If you have the storage space and the money upfront, it is worth arranging to get a year's supply at harvest time rather than buying small amounts throughout the year. Many farmers will run out before the next lot is ready, and it can be stressful going on a wild goose chase trying to find another supplier. During our time living in rental properties with goats, we always just bought two to three weeks of supply at a time. We've had to put up with very expensive hay at certain times of the year and would have preferred to avoid this, but at least our goats got fed.

The amount of hay that your goats will eat depends on whether they have other food. If you're feeding them hay only, without grazing, two adult goats will generally go through between one and two small rectangular bales every week, usually around three bales every two weeks. This depends on the quality of the hay and how tightly it has been baled.

The dry weight of food that a goat will eat is estimated to be between 3.5 and 5 percent of their body weight each day, so for a 143 lb. (65 kg) goat, this works out to be between 5 lb. and 7.1 lb. (2.27 kg and 3.25 kg) of dry food per day. All food has some moisture content, so its actual weight will be higher than the dry weight. If your goat eats a lot of scraps, fresh pasture, and leaves, it will work out to be more in weight than if she was only eating Lucerne. If the food they're eating is low in nutrients, they may eat more, and if it is nutrient-dense, they will eat less. Observation is always best—if your goat seems hungry, then he/she probably is. It's always best to allow free choice access to staple foods such as Lucerne or tree branches. Goats will adjust their own food intake depending on their energy and nutrient needs.

Goats are ruminants, which means they have four stomachs, one of which contains bacteria that ferment their food to digest it. Because of this bacterium, it's important to allow goats time to adapt to any new feed given to them. Feed them garden scraps in moderation. Upsetting the goat's digestion by introducing too much new food at once can lead to serious health problems, even death.

Feeding Scraps

Goats enjoy most fruit and vegetable scraps such as apple cores, orange peels, banana skins, the limp outer leaves of cabbage, broccoli stems, pumpkin skin, and vegetables we don't eat. They also enjoy scraps from bread and other baked goods. Goats should be fed no meat, nor anything poisonous such as potatoes that have gone green or anything moldy. Individual goats seem to have different preferences for scraps. One goat I look after thinks that banana peels are the best things ever, but the other goats won't touch them.

Feeding Trees

Goats love to eat tree branches and leaves. If you have access to suitable trees, you may be able to get away with not buying any hay. Goats generally love nitrogen-fixing trees like acacias and tagasaste. They are fond of most maple leaves (although red maple is poisonous), tree ferns, other ferns species (not bracken though), willows, apple trees, and pear trees. They like ash, elm, oak, poplars, and pines.

Goats have a good sense of what they can and can't eat, so if you're unsure about whether something is a suitable food for them, you can give them a small amount, with plenty of their usual food to eat. This way, they are not forced into eating only the new stuff, and you can see their reaction to the new food. I have found that goats prefer different trees at different times of the year, and some prefer different plants to other goats.

In earlier times, "tree hay" was often made in the summer from ash, elm, ivy, and oak. If you have storage space, you can cut small branches off these and other goat fodder trees and dry them in bundles hanging from the rafters. Nettles and other leafy plants can also be treated in this way. If the tree hay branches you collect are thin enough (around 1 cm - 1/2" thick), goats will often eat the whole thing, branch and all.

Goat fodder trees can often be found on public land, so if you don't have many trees at your own house, you can always go for a daily stroll with secateurs or a pruning saw to gather branches.

Poisonous Plants

Rhododendron and azalea are highly toxic to goats. Many other garden ornamentals are a bit suspicious as well, so make sure that before you feed anything to your goats, you've looked it up first to check it's not toxic to them.

There are lists online of plants that can be toxic in high enough doses, but goats will often eat small amounts of these with no problems. The key to avoiding poisoning is always to have plenty of food accessible that they will eat. Before feeding any new plant, make sure you've identified it and that it's safe for goats. Before tethering a goat, check that there's no rhododendron in reach, that bracken ferns have been thoroughly stomped down or removed, and that there's plenty of food within reach that the goats are eating at this time of the year.

Plum, peach, nectarine, and cherry leaves can be toxic to goats, so try to make sure you don't have these trees anywhere near your goat paddock. Many believe that the leaves of other plants aren't good for goats, but from my experience with sycamore maples, as long as there's plenty of other food for them, it's not a problem.

Growing Food for Goats

In a backyard, it is best to either keep plants separate from goats and bring small amounts to them or to offer controlled grazing, either by tethering them nearby for short periods or by growing plants against the outside of their fence so a few leaves can be accessed from the goat paddock, but the goats can't gobble the entire plant up or eat all the bark off the trees.

Goats appreciate access to comfrey, either a couple of leaves offered in their feed each day or being tethered near comfrey for a short time (with plenty of access to their other favorite plants in the same place).

Roses are good as a remedy for scouring, and goats also appreciate the taste of them.

A variety of kitchen herbs and "weeds" can be grown and offered to the goats with other feeds. Goats might choose to eat at certain times and not others, but a variety of food is good for their health, so it's worth growing a few extra herbs to share.

Minerals and Supplements

Copper is the most important mineral to add to goat diets in areas where it is deficient in the soil or where the water is high in sulfur, iron, or calcium. Darker colored goats have a higher need for copper than white goats, and you can often tell when a darker goat is deficient in this, as their coat will become lighter. Loss of hair on the tip of the tail to give it a "fishtail" appearance is another sign of copper deficiency. In *Natural Goat Care*, Pat Coleby states that she's never encountered a goat with worm problems when their diets have been supplemented with copper. Copper sulfate can be bought in animal feed stores, and it's often found with horse supplements. The easiest way to add it to the diet for milking animals is to mix one teaspoon per goat into the "treat" rations each goat receives. Another way is to place it in small containers available to the goats all the time, but make sure they are kept indoors, or they will be ruined every time it rains.

Copper oxide is more difficult to find than copper sulfate but is a safer option for those worried about copper toxicity.

I had never heard or read anything saying it is possible to feed too much copper to your goats until recently, and there is still no definite upper limit for it. I found when I was feeding my Toggenburgs copper sulfate every day (around half a teaspoon a day—over three times more than what I recommended above), they were very healthy and had no parasite problems. Here in Australia, the soil is often low in lime minerals and copper, and our goats benefit from the extra copper, but if you have healthier soil, keep to 1 teaspoon a week for copper sulfate or to feed them copper oxide instead.

If sulfur is deficient in your soil or notice skin problems or external parasites on the goats, it might be worth offering gypsum or yellow dusting sulfur to them. You can do this as a free-choice mineral or sprinkle it into their food.

Kelp (seaweed) is an excellent natural supplement that supplies a wide range of minerals, especially iodine, which is essential for absorbing all other minerals and vitamins and especially important if you're feeding your goat's Lucerne. Kelp is best-given free-choice, either by having a container on a wall out of the rain for the goats to eat as they choose to or offering it to the goats at milking time twice a month to see if they are interested. I like to sprinkle a small amount on top of their treat-feed every day. Sometimes they will eat large amounts of it, and other times they are either not interested or will only eat only a small amount. If they have never had kelp before, you may find they eat a lot at first.

Selenium is an important mineral for goats. Kelp, wheat, oats, and sunflower seeds are good sources of this (as long as the soil they're grown in is not deficient), so your goats may already get the right amount via their milk treat. If you're concerned about the mineral levels of any goats that don't get grain, give them a handful of sunflower seeds regularly. Sulfur is needed to absorb the right amount of selenium, so supplementing with Sulfur if the soil is deficient or acidic is a good idea.

Salt is essential for goats if you are not feeding kelp, but you may find they get enough of it from their regular feed. Goats know when they need salt and when they don't, it's best to offer it free choice (preferably as kelp).

Goats will sometimes eat a lot of salt when they need potassium, so adding cider vinegar to their water can add extra potassium to the diet. Goats can be fussy about licking from a block of salt that another goat has been licking, so loose salt is preferred to blocks, or get one block for each goat. I have used Himalayan salt, but any natural unrefined salt without additives will do the trick. Instead of having a salt lick for the goats to lick when they want, you can give them handfuls of coarse salt to nibble at now and then, either out of your hand or a bowl.

Calcium and magnesium are very important for dairy goats. Dolomite lime, either offered free choice or around a tablespoon per day added to the treated feed, will supply both of these minerals.

Potassium is important for pregnant does as kidding time approaches. Apple cider vinegar added to the drinking water is a good source of this and is an excellent supplement throughout the year to boost immunity and digestion.

Pregnant and lactating goats need a supplement, too. Usually, this is a mixture of locally grown grain. Barley is said to be especially good for dairy animals, as it increases the amount of milk. This supplement is best fed as a treat at the milking stand. Feeding a pregnant goat like this every day will get her used to coming to the milking stand and will make milking a lot easier once she has kidded. Avoid goat pellets at all costs. Goat pellets turn to something like mushy cardboard inside the goat's belly, and goats need more fiber than pellets provide. Also, try to avoid anything with molasses in it or sweetened, as sweet foods make them more prone to insect attacks. To buy prepackaged "treat" feed, dairy meals designed for cows, made from cracked and rolled grains, can be an acceptable choice, or just buy a large bag of whole wheat and a smaller bag of sunflower seeds and mix them together, or even just plain wheat or plain barley is good. Soaking whole grains in water with a splash of cider vinegar overnight or for 24 hours will enhance their nutrient availability and make them more digestible. I usually soak one batch of barley in the morning each day. Some of it is fed in the evening, and the rest of it the next morning. Before feeding the barley, I drain the soaking water and then mix in the daily rations of copper sulfate, yellow sulfur, and dolomite lime into the grain.

Before you get your goats, research soil mineral deficiencies in the area you'll be buying hay from and try to offer these minerals as a free choice. Alternatively, carefully add small amounts to their treat feed. Offering minerals as a free choice makes it easier for the goat to correct her own nutrition when she needs to, but you'll need a way of keeping these out of the rain. If the goat isn't interested in the minerals, you can try sprinkling a little grain over the top to encourage her.

For Australia, where our soils are mostly acidic and deficient in copper, Pat Coleby recommends a basic stock lick made from 12 lb. (6 kg) of dolomite, 2 lb. (1 kg) yellow dusting Sulfur, 2 lb. (1 kg) copper sulfate, and 2 lb. (1 kg) kelp. These minerals can be found in animal feed shops and horse supply stores. Dolomite is easily found in any garden center.

Hoof Trimming

Goats are from mountainous areas where their hooves are worn down from daily wandering and jumping on rocky ground. We can imitate this to a certain degree by having large rocks for the goats to climb on in their paddock, but you still need to keep an eye on their hooves, which generally need to be trimmed every eight weeks. Specialty hoof trimming shears (also called footrot shears) designed for sheep or goats are the best tools for this job, but garden sheers can also be used, or a sharp knife if you have enough confidence and a goat that stays still. If you don't trim their hooves on time, the hooves can grow long and curl around over the base of the foot, trapping mud and goat poo, which may rot and cause health problems.

When trimming a hoof, it's better to start by trimming a small, even slice all around the hoof first, then trimming another small amount until you are very close to the foot. Everything should appear to be even, clean, and comfortable. It's possible to trim too closely, and the goat can get cuts on their skin from doing this, so it's better to try a little at a time, and if there's any doubt about whether you've trimmed enough off or not, it's better to err on the side of caution and not trim off anymore. It's good to do this on the milking stand with a bowl of treats for the goat to eat. Depending on the personality of your goat, you may need someone to help hold its leg still. You may need to do the hoof trimming over two or four days, especially if you've been milking the goat on the stand before you start, as they might decide they've eaten enough treats and that it's time to go back to the paddock.

Milking Stands

This isn't essential right away, but it will make life a lot easier when you need to milk your goats or trim their hooves. There are free instructions available online for making them out of pallets and other wood. They can occasionally be picked up second-hand from sellers on Craigslist, Gumtree, and other classifieds. A good milking stand will have a means of securing the goat to the stand, usually by having her head go through an opening that can be closed into a size big enough to be comfortable around her neck but small enough she can't move her head back through. Another way is to have the goat on a leash secured to the milking stand or a wall beside it.

Water

Goats need clean water available at all times, and for a small backyard herd, this is easy to manage. I've found it easiest to provide this in sixteen-liter (three-gallon) buckets. I use one or two in winter and three or four on hot summer days. It's important to check up on it twice a day to make sure they haven't drunk it all or that it hasn't frozen in winter. We attach one of these buckets to a clip at the end of a rope attached to the fence. My husband can reach over the fence to lift it up and down using the rope so we can easily refill the water without going in and out of the paddock. Refilling the buckets via a watering can is another quick option, but you must still remove the buckets when they need to be cleaned. Rubber or flexible plastic tubs designed for horses work well for goats. Cheap plastic buckets can be used but will not last long.

Always check to make sure your goats haven't defecated in the water. If they have, it should be changed right away. Try to keep a couple of buckets in different places in the paddock. It's less likely the goats will knock them all over or defecate in them simultaneously.

If you live in a climate with very cold winters, you will need to either insulate the water container or use a heater system.

Goats appreciate warm water in winter and cool water in summer. You can put ice cubes in the water on very hot days, and on cold days, they will appreciate a bucket of warm water if you can manage it.

Chapter 6: Milking Your Goats

How to Milk a Goat

Before you begin, make sure you have everything ready you need for the milking and straining. The straining cloth should be boiled, the jars and funnel sterilized (see the next section for information about doing this). Fill a food bowl with your treat feed and place it in the milking stand's feed bucket area.

The amount of treat feed to give a goat will depend on how much milk she gives. Some goats will handle higher amounts of grain than others. Some goats produce more milk with higher amounts of grain, while other goats seem to do better with less grain. Approximately two cups of grain is a good amount to start with. Have your milking bucket close by, but not anywhere that the goat can easily knock over or anywhere else that it can easily be knocked over or contaminated. I keep a small table near the milking stand for the bucket and jars.

Walk the goat up to the milking stand, guide her head through the headgate and secure it around her neck. You will now need to clean her udder. Either brush it with a dry cloth or the back of your hand to remove stray hairs and dirt, or if she's very messy, you can wash her udder with a wet clot. Rub it gently with a very dry towel, making sure she is dry, as you're far more likely to get sick from dirty water dripping into the milking bucket than you are from a few stray hairs or bits of dirt.

If you're washing and drying the udder, you will need to use a separate cloth for each goat. Few goat books recommend following the first method (my preferred one) for udder cleaning and prefer the more thorough washing and drying approach. A lot of the belief in washing goat udders comes from milking cows, as they seem to be attracted to the muddiest part of the paddock and can have very dirty udders. On the other hand, goats will find the driest place possible and don't seem to get dirty often. The simple method works for my family, as we keep our goats clean and dry with lots of straw, but if your goats are covered in muck, washing and drying is the best option. My mention of the simple method might be a controversial approach, but it is far easier to brush the udder quickly than to wash and dry the udder. If you're in doubt, or can't chill the milk quickly, or are sensitive to food contamination, wash and dry the udder, making sure you dry it thoroughly.

Make sure you are seated comfortably. I sit on the edge of the milking stand, but plenty of people use stools instead. You should be able to sit there milking the goat with no need to bend your back and without stretching your arms out awkwardly to reach the teats.

Take two squirts of milk from each teat, milk it onto the milking stand, ground, or a separate dish. By discarding the first squirts of milk from each teat, you decrease the risk of contamination from anything lurking on them. Once you have discarded these first squirts and your goat's udder is clean, place the milking bucket close to the udder and begin milking into it.

To milk a full-sized goat, first place your thumb and index finger around the top of the teat where it meets the udder and close it off, then close your middle and ring fingers (or just the middle finger if her teats are small) around the teat to squeeze the milk out of it. Repeat this with one hand after the other until it becomes more difficult to get the milk out of the teat. Remove the bucket, and then massage her udder or mimic the action that a kid uses on it by pushing against it with your hand, then place the bucket back under her and continue milking as you were before. This helps her to let down as much milk as possible.

If your hands get tired using that method of milking, you can alternate with another method where the thumb is not used, and just the index finger is used to close off the top of the udder. This method uses different muscles to the usual milking method but can't usually be done until the udder has emptied a bit.

To get the last of the milk, "strip" the udder using both hands on one half of it at a time. Gently squeeze the milk from that half of the udder into the teat and then out into the bucket. When it's time to stop milking, she will be giving the tiniest milk (or none at all).

It's easiest to watch someone else milking to learn and if you don't have anyone nearby, try searching for videos online. To get your hands used to the action of milking, go through the motions of milking on your thumb. It's a good idea to learn to milk while the kids are still drinking their mother's milk because the kids can help drink the rest of the milk, and your doe is not at risk of udder problems or drying up as long as someone is taking the milk. Once you are in a good milking routine with your goat, milking will take less time than straining and cleaning. I take around five minutes to milk a goat with a good udder – a bit longer if her milk is slower to flow.

When you've finished milking your goat, leave her on the stand to finish her meal while you strain the milk. This extra time on the milking stand helps the teat close before any bad bacteria can get into it. To strain the milk, place a funnel over the top of a glass jar, then cover the funnel with a sterilized thin cloth, such as butter muslin or cheesecloth, and pour in enough milk to fill the funnel. If it's taking a long time to go through the cloth, you may need to find a thinner or more loosely woven cloth for next time. Alternatively, gather the edges of the cloth in your hands and tilt them around carefully to get more milk through. The creamier the milk, the longer it will take to strain.

A Milking Routine

- Boil the straining cloth (or boil it the night before)

- Assemble everything you need during and after milking (e.g., clean jars and funnel on the table, clean milk bucket near the milking stand)

- Bring the goat to the stand, clean her udder, milk her

- Strain the milk

- Take the goat back to her paddock

- Repeat for other goats

- When you've finished milking all the goats, wash the straining cloth and hang it up to dry, sterilize the bucket and funnel

Milking Equipment and Hygiene

When the milk is in the udder, it is sterile and is a perfect food for baby goats to drink. You rarely hear of baby goats getting sick from their mother's raw milk in the same way that the media sensationalizes cases of humans getting sick after drinking raw milk. Due to its neutral acidity, any milk out of the udder, raw or pasteurized, is an easy medium for bacteria to grow. This can be beneficial when we encourage good bacteria to make cheese, yogurt, and other fermented foods, or it can be bad. If you're drinking the milk right away, hygiene is not much of an issue. The longer you wish to store it, the more careful you need to be about the storage conditions and the likelihood of anything coming into contact with the raw milk.

There are two main approaches to preventing contamination by the wrong bacteria on milking equipment. The most sustainable and healthy option is heat sterilization. Any containers used for the milk must be heated to a safe temperature first, then the milk can be added. The other main method is with chemicals.

After the equipment has been sterilized, it needs to be kept away from anything that could contaminate it. I recommend milking buckets with lids. That way, you can keep the funnel and the bucket's interior away from potential contaminants. I use a bucket with a 7 quart (7 liters), but around half this size is sufficient. Empty the bucket as soon as you've milked each goat to avoid it being kicked or stepped and losing the milk. Buckets larger than seven liters will be difficult to fit underneath the goat, so search for a small stainless-steel bucket with a lid. It's tricky to find, but it's something that only needs to be purchased once, so it's worthwhile getting the right one to start with. A stainless-steel stockpot can be used instead of a bucket. Try to find one without air holes in the lid, and you will have something that works just as well as a bucket with a lid.

Heat sterilization can either be done with boiling water or in the oven. Using boiling water can cause injuries (and rude words), so the oven is preferable. To use the oven, everything you put in it should be metal or glass. Plastic can dry after being boiled once the oven has been turned off, but anything put in for long enough to sterilize it should be able to withstand the heat, so I recommend stainless steel milking buckets, stainless steel funnels, and glass jars.

To sterilize using the oven, place your clean bucket and jars in it upside down, and switch it on to 230ºF (110ºC). Don't use ovens with visible (orange-glowing) electric elements for sterilizing glass, as the glass can crack. Leave the oven to heat up, and once it's been fully hot for at least five or ten minutes, test it by touching something sterilizing in there—it should be very hot to the touch. Turn it off, close the door, and leave it to cool down. First, jar lids and plastic funnels should be submerged in boiling water for 30 seconds, and then the water drained off. Place them in the oven as it cools down to dry them and keep them sterile until you need to use them. Be careful that your plastic will stand up to this as some plastics are flimsier than others, but the plastic Ball mason jar lids will without any problem, as do good funnels designed for jam.

For the straining cloth, I rinse it and hang it out to dry after every milking, and then boil it when we're ready to begin milking the next time.

If you need to sterilize everything with boiling water, boil enough so you have plenty to pour over everything that needs to be sterilized. The aim is to heat the jar, funnel and bucket surfaces to kill off any potential nasties that might be lurking there, so slowly pour plenty of boiling water over all the surfaces.

The jars should be wet (but with no standing water) before adding boiling water so they don't break. Pour the water into the jars, filling them to around one-third of the way, and put the lids on. Turn the jars on their sides and rotate until the water heats the jars and you can't comfortably handle them. These days I rely on a wood cooking stove that's often slow to boil water in the mornings, so at night I put the straining cloth in the bucket with the funnel and leave it all submerged in the boiling water overnight, draining it all in the morning.

Every time you empty a jar of milk, rinse it twice with a little cold or lukewarm water and leave it until you're next ready to sterilize jars. Leaving jars sitting with a little milk in them makes it hard to remove it later.

Handling and Storing Milk

Once the milk is in the jars (if you're not drinking it right away), you should chill it quickly. I surround each jar of milk with ice-filled containers (I use the "ice bricks" made for coolers). You can even chill the jars before you milk, so the cold jar begins chilling the milk as soon as it is strained. After the milk has been chilled, it's important to keep it at a low temperature and not to let it fluctuate. Many people keep their milk in the refrigerator door, but this is not a good place for longer-term milk storage. Only store it there if you will be using it in the next 24 hours. The temperature fluctuates too much in the fridge door, and it's also the warmest part of the fridge. It is best if the milk is on a shelf in the fridge, preferably toward the back. Avoid leaving the milk sitting out at room temperature for entire mealtimes or similar periods unless you plan on drinking the rest of the jar soon.

Don't place warm containers of leftovers or anything else next to the milk. If you're adding a lot of room temperature and warmer things to the fridge at once, add ice-filled containers right next to the milk to make sure that it stays cold.

The quicker you'll be drinking the milk, the more relaxed you can be about its storage. Bacteria need both the right temperature and the right amount of time to multiply to dangerous levels.

Now that I rely on a small off-grid system without a fridge, I find that in winter, we drink all the milk within around 24 hours, so the temperature of an unheated room out of the sun (50°F/10°C) is cold enough for it, even without ice bricks. In summer, there is a lot more milk to handle. On the hottest days, anything that doesn't get purposely fermented into cheese or yogurt begins to ferment on its own within 24 hours unless I'm super careful about switching the ice bricks around.

1-quart (1-liter) mason jars and 22 ounces (650 ml) Passata bottles are good sizes for storing fresh goats' milk. To make the easiest "chèvre" from milk warm from the udder, it is worth having a 2-quart (2-liter) jar for this purpose.

Excess Milk

If you're lucky in your search for goats, and the goats you thought would only give one liter (one quart) a day each instead give two liters or more, you may find you have more milk than you can use for drinking and cheese-making. Depending on local laws, selling or giving away the excess milk is one option. A herd-share arrangement may also work. The legalities of selling raw milk vary depending on the country and state, so check before you sell.

Another option is to feed the milk to animals. Pigs can drink the milk as it is. While chickens will drink the milk, they will gobble it up in its solid form far more readily if you make "ricotta." Milk left sitting at room temperature for too long will often begin to turn to cheese on its own and can be fed to animals.

Goat's milk is great for the skin. You can either add it to baths or dilute it and use it on your skin. Fresh goat's milk applied to a sunburn and other skin conditions can be soothing and healing. Goat's milk soap can also be made.

Diluted raw milk sprayed onto pasture or garden plants works as a great anti-fungal and fertilizer. This is my way of making lemonade from lemons when a goat puts her foot in the bucket—at least the garden is getting a drink.

Excess whey from cheese making can be fed to pigs and chickens and made into whey cheeses such as gjetost. Whey with live cultures from chèvre and hard cheese (not acid-curdled whey from ricotta or paneer) is great for adding as a starter culture to fermented vegetables like sauerkraut and pickles. Whey can also be used instead of water in stock and bread, as soaking water for grains, and as cooking water for vegetables, grains, and pasta. Whey from cultured cheeses (not acid-curdled ones) can be diluted (one-part whey to nine parts water) and used to water plants.

Chapter 7: Goat Grooming, Health, and Hygiene

Keeping a goat healthy requires you to keep its digestive system healthy. A goat's diet is plant-based because they are ruminants. The goat has three fore-stomachs, so each has an omasum, rumen, and reticulum. The true stomach is the fourth one and has the name abomasum. When the goat eats hay, it is digested in the forestomach. The abomasum helps break down the protein, carbohydrates, and fats, and it resembles the human stomach in this way.

Parts of the Digestive System

Every part of the stomach does unique work. Among the forestomach, the rumen is the largest, with a capacity of one to two gallons. It has bacteria that help ferment the hay and break it down. This hay is regurgitated, chewed, and swallowed again. In the process, methane gas is produced. Rumination causes belching, which has a strong odor indicative of a healthy rumen. There is heat produced in this process that helps keep the goat warm.

The second part of the forestomach is the reticulum, and this is close to the liver. This stomach compartment works with the rumen and helps break down the food the goat eats into smaller units to absorb and use for its blood circulation, breathing, and other metabolic processes. All bigger pieces of food get stored here. The reticulum pushes the easy-to-digest food to the mouth for chewing as *cud*.

Chewing and Digestion

This chewing and regurgitating process will continue until all the food is small enough to comfortably enter the omasum. The food is broken down even more by the action of enzymes in the omasum. The abomasum is the only place where the digestion of the food items such as milk and grain occurs, as they do not need bacteria. The broken-down food is sent to the intestine, where the useful material is absorbed and the waste eliminated.

Other Parts of the Goat

Hooves

One has to understand the role that hooves play for the goat. Hooves play an important part in the movement of the animal. If the hooves are injured, it affects all the other parts. The goat will feel pain, will limp about, and if not rectified, it can lead to a shorter lifespan. Care for the hooves will keep the goat happy. Trim the hooves to avoid scald and rot since it could lead to death. The hooves help the goat climb the high, steep, rocky terrain. Spanish goats have a feral nature, and so they need little care for their hooves. A good, well-cared-for hoof will have the shape of a rhomboid.

Teeth

The goat's teeth differ greatly from those of other animals. It has a hard pad on the top of the mouth in the front. There are no teeth there; it has teeth only at the bottom. There are teeth on top and bottom in the back of the mouth used to chew the cud. Like all other animals, baby teeth fall out with age. They grow adult teeth, and this helps establish whether the goat is an adult or not. Vets check for the growth of the teeth on the lower jaw in the front of the mouth. By the time they are five years old, the goat must have all eight adult teeth in place.

Beard and Pupil

Several varieties of goats have beards that are more prominent among males. Sheep do not have this beard. Goats use this beard during mating to attract the female because the beard has a scent. In dairy goats and pygmy goats, we find the wattle, which is attached to the neck. The eyes of the goat can be confusing because they have many shapes and colors. The main difference is that goats have a square-shaped pupil instead of a circular one.

This peculiar shape of the pupil helps goats see in the dark, which is important because goats need to stay away from predators. The color of the eyes ranges from blue to brown and yellow. Angora goats have hair over their eyes, and this hampers their ability to see properly.

Difference Between a Sheep and a Goat

You can spot a sheep from far off because the tail will stay down while that of the goat will stay up. Also, the horns of the goat will be straight, while sheep have curved horns. Sheep will not have beards, but goats may have a beard. While grazing, the goat tries to be independent, and the sheep will try to follow others.

Visual Inspection

You can tell a healthy goat by the shine in their eyes and their coat. They will be playful if they are healthy. They will play around until it is time to chew their cud or sleep. They will have good posture, standing tall with tails, head, and ears upright. If the goat is sick, it will have a droopy posture. The sick ones try to relieve pain in the stomach and other parts of the body by stretching. If a goat is trying to stretch or urinate excessively, it is a sign it is not well.

Keep a Record of Good Health and Behavior

To keep up with your goat's health, know how your goat behaves when it is in peak condition. Observe how it jumps around and interacts. If there is any deviation from this behavior, you must act. Although you do not have to respond to their every call because goats make lots of sounds, tend to them when they ask for food or water. If you make the feeding time fixed, it will help you understand their cry perfectly.

And, if your goat is not feeling well, it will moan or keep silent but will not cry out. The doe may cry when they are in heat, while bucks in a rut will make peculiar noises. It does make noise while giving birth. The sound is more like a whine than a cry, but the cry will become distinct and loud when it is pushing time. The cry of a goat trapped or in pain will be frantic and plaintive. You can give them enough care once you recognize the different types of cries.

Keep the Goat Warm

The goat's internal temperature is around 102°F, although there might be a one- or two-degree difference from one goat to another. On hot days, their temperature will be higher. If there is a profound change (more than one degree) in their temperatures, it is a sign of illness. When a goat has issues with the rumen, it will show a low temperature. You must warm it, or it will die. To do this, buy a coat made for goats to keep them warm. If you cannot get one, make one with human sweaters or other clothes.

When you are in charge of goats, record their normal behavior and state. Take the goat's temperature throughout the day and note this along with the time. You can compare the readings in the future using this. Check the type of thermometer you use; either a digital or glass thermometer is okay. Always shake the glass thermometer well before you use it. Tie the thermometer with a piece of string so you can pull it out if it goes in too far.

Taking the Temperature of a Goat

To get the temperature of the goat, hold the goat still. If it is a fully-grown goat, you may need a stanchion to hold it. Use petroleum jelly to lubricate the thermometer and then place the thermometer inside the rectum. Keep it there for two minutes. Remove it slowly and note the temperature. Clean and sanitize the thermometer using alcohol. Cud chewing is normal, and every healthy goat must be doing this (ruminating). So, watch your goat and see whether it is ruminating. You can check the rumination by looking at the abdomen of the goat. Listen on the left side, and you will hear a growling sound along with a bit of movement once every two minutes. You may also use a stethoscope to listen to this sound.

Measuring the Pulse and Respiration

Goats rest during the early afternoon period. This is ruminating time for them, and if you find any who aren't, then check them to see if they are unwell. The easy way to do this is to discover their pulse rate. Ordinarily, they should have a pulse rate between 72 and 85 beats per minute. Kids will have twice this rate. Before taking the pulse, make sure the goat is resting. Place your finger below the jaw, and you will feel a pulse. Take the count for 10 or 15 seconds and multiply the number by 6 or 4 to get the beats per minute.

It is possible to discover how healthy your goat is by measuring its respiration rate. For an adult goat, this will be between 12 and 30, whereas, for a kid, it will be between 22 and 38. You can measure this easily by looking at the goat while it's resting and counting the number of times its side rises and falls. By the time the goat is two years old, it is almost fully grown. They continue to grow until they are three, but the growth won't be so rapid then. On average, all goats live between 7 and 12 years, but they can live much longer.

Wethers have a longer life expectancy, and people often raise them as pets. The main cause of death here is urinary calculi. For the does, death is more often due to kidding problems. It is customary to stop kidding after they turn ten because this will lengthen their life span, and they may even live another ten years or more.

Preventing the Onset of Disease

Most diseases come with newly purchased animals. This may come through the clothing and footwear of the people handling them. Other pets, birds, and mice might also bring in germs that cause disease. To prevent this, one has to clean the shelter, vehicles, footwear, and mats with a bleach or disinfectant such as Virkon S. You could also use chlorhexidine, an antiseptic.

The deficiencies of nutrients in the diet, including copper, selenium, calcium, zinc, or protein, could lead to poor health. The breeders must check the soil to make sure that all nutrients are present and plentiful in the pastureland.

Keep an Eye Out for Symptoms

When goats sneeze, cough, or look lethargic and do not keep up with the herd, it means something is not right with them. This indication gives you a warning 3-4 days before something happens. Use a stethoscope to check the heart rate and measure respiration. If you check your goats' health daily, do it at the same time.

Major Diseases

It can be hard to diagnose a disease because many have the same symptoms. It is better to work with veterinarians and use their treatment recommendations.

Caprine Arthritis Encephalitis

This viral infection affects the lungs, joints, mammary glands, and brains, which will spread to the colostrum once the goat is infected.

The signs and symptoms will not all show up at once. Test the bucks and the semen because it can spread to the does.

Caseous Lymphadenitis

This is a contagious disease spread by bacteria through the lymph glands. You can control this disease by culling.

Contagious Ecthyma

This is the sore mouth disease you can see on the lips of kids. It spreads through contact, and scabs will appear on the vulva, teats, scrotum, face, and lips. Lesions will disappear in two to three weeks. There is a chance for the animals to get a secondary infection. Wear gloves since this is a zoonotic disease.

Footrot

This is another contagious bacterial infection that occurs in the soft tissue of the toes. You will see swelling, pus, and redness in the toes of the goat. It has many forms, so it is better to have a swab analyzed at a laboratory.

It is necessary to trim the hoof when needed. Give the goats high-quality zinc with the mineral mix. To help prevent footrot, encourage the goats to walk on rough surfaces to make their hooves stronger. Also, pay attention to the feet of the animals during selection.

Listeriosis

This bacterial infection causes paralysis of the trigeminal and facial nerves with discoloration of the eyes, fever, and depression. This bacterium will last for years. Infection occurs when the goats feed on silage. You can treat it with off-label drugs.

Overeating Disease

A bacterium named clostridium perfringens causes overeating disease. This bacterium is found in the soil and the intestinal tract. When the goat is not acclimatized to the feeding land of the pastures or wanders into bushland with fast-growing cereal crops, this increases the chance of infection.

You can prevent the spread of this disease by vaccinating the doe three weeks before they kid, so the bacteria is not passed on through the colostrum. Vaccinate the kids six weeks after this.

If the goat gets infected, it will stay sick for several weeks, while the kids may die before the symptoms are even noticed. They will have no appetite and will have intermittent diarrhea.

Pneumonia

Pneumonia manifests in many forms in goats. This is linked to fungi, bacteria, parasites, or viruses. Before you begin the treatment, it is important to know which form of the disease you are dealing with.

Goat Polio

This is a Vitamin B1 deficiency that can be cured by giving your goat thiamine (Vitamin B1). The symptoms appear if there is any change in the diet of the goat. This includes moldy grain or hay, improperly formulated rations, sudden feed changes, and giving molasses-based grain that has mold. If you don't treat the goat immediately, it will die within 1-3 days.

Leptospirosis

This abortion disease can be avoided by vaccinating before you begin breeding. Since this disease is contagious, you have to wear gloves when handling dead fetuses.

Urolithiasis

When stones form in the male urethra, it causes urine retention. This condition is called urolithiasis. It comes with abdominal pain and may cause bladder rupture. There is no specific reason for this, as it is not dependent on season or feed. The bucks sometimes break their penises while breeding. This condition is not treatable.

Bloat

This occurs when you feed the goat pellets, milkweeds, hay that is wet and moldy, or alfalfa straight. The goats cannot burp, and this causes serious problems. At times, this problem is caused due to an obstruction in the esophagus. Offer baking soda to the goat to regulate the working of its rumen. This will help remove the gas naturally.

White Muscle Disease

You will see sudden trembling and stiffness when white muscle disease afflicts the goat. When the pastureland doesn't have enough selenium, it can lead to this disease. Increase the Vitamin E and selenium supplement in the mineral mix to overcome this problem. Kids must be given a shot as soon as they are born if the pasture is deficient in selenium.

Scrapie

It is a form of degenerative disease fatal to goats. It attacks the nervous system and is a form of transmissible spongiform encephalopathies. The deadly mad cow disease can be possibly transmitted to humans through this Scrapie disease. Remove afterbirths fast, then clean and disinfect buildings where kidding takes place.

Infections by Parasites

When parasites infect your goats, they will show decreased appetite and loss of weight. Common parasites include the Brown Stomach worm and the Barber Pole worm. Use a chemical de-wormer if you think the goats have worms. Check the insides of the eyes.

Additional parasites include Razor worms and Brown Stomach worms. They reduce appetite, cause diarrhea and weight loss. Compare the color of the inside of the eye with the FAMACHA scorecard, and it should be pink.

What is the FAMACHA Score Card?

FAMCHA is derived from **FA**ffa **MA**lan **CHA**rt and is a method used in South Africa by goat breeders when they spotted that, although they were regularly worming their herds, the worms were not disappearing, but instead getting worse. They discovered that the worms were building up resistance to the chemical de-wormers that the herders were giving their goats constantly. Once the worms were resistant, there was no way to get rid of them, and the result was the gradual loss of their goats.

So, what do we use FAMACHA for? It is used to determine parasites in sheep and goats and to detect anemia caused by them. A scorecard is used, with a scale of 1 to 5 – lower scores indicate fewer anemias, which means less risk of parasites. Always use this scorecard – if you try to score by memory, you risk serious issues if you miss-score an animal.

Your local vet should be able to provide you with the FAMACHA scoring cards. You can also get in touch with the American Consortium for Small Ruminant Parasite Control to learn more about the FAMACHA scoring training.

How to Use the Scoring System

This will require a scorecard, another person to help you, and a holding stand.

You need to check the membranes of your goat's eyes in natural lighting – sunlight gives the best chance of accuracy. Also, make sure that your shadow doesn't shield the eye you are looking at.

A method to remember how to check the membrane is this:

- **COVER** – the eye must be covered with the top eyelid
- **PUSH** – push the eyeball – not hard, just enough that the top eyelid's eyelashes curl backward over your finger
- **PULL** – pull the lower eyelid down. If you do this right, you should see the inner membrane

Now, quickly check the membrane's color, looking at the pinkest part against the FAMACHA scorecard. You must do this quickly because if you take too long, the goat's eyelid will dry out, and the membrane will turn red.

So, what do these scores mean? The scores are from 1 to 5. The lower the number, the less chance of parasites. If your goat scores 1 or 2, there's no need to worry about worming right now. However, a score of 4 or more needs immediate treatment because not doing so could lead to death. If your goat scores 3, you need to make a judgment call on whether to worm them or not. If you have a few goats and they all score 3, it may be best to worm them to stop the problem from worsening.

Some people think it's easier to forget about the FAMACHA scoring and just worm their goats anyway; this can be dangerous, as we explained earlier, because the worms develop resistance. If you are at all unsure, consult your vet.

Chapter 8: Goat Breeding and Kidding

If you hadn't realized yet, the only way to get milk out of a goat is to breed her and let her have babies. She will produce no milk until that has happened, and the process is known as freshening. To succeed at breeding your goats, you might be interested to know these facts about a goat's breeding cycle:

Bucks, or male goats, can start breeding from as young as seven weeks old. This does not mean you should allow it, just that they are sexually active and can get their sister or mother pregnant. Separate the bucks away from the does before they are seven weeks old.

Bucks can breed at any time and will go 24/7 if allowed, unless there are extreme weather conditions. They will not breed for fun, though, only when they can smell that a female is in heat.

Bucks go into a "rut." This means they get a real surge of hormones and are ready to breed before the female is ready. Occasionally, when a buck goes into a rut, it can make the does go into heat. During the rut, bucks can be very dominating and will do crazy things, some of which will make you laugh, others that will make you cringe! They will snort, spit, and urinate on themselves to make themselves smell worse and may even drink their own urine.

A doe goes into heat on a 21-day cycle, and each heat lasts between 1 and 3 days. There are breeds, like the Nubians, Spanish, Boer, Fainting, Pygmies, and Nigerians, that can breed all year, but most dairy goats are seasonal. This means they will only go into heat in the fall, between August and January.

A full-size goat can breed at about eight months or when they reach 80 lbs. in weight. Try for the year mark before allowing them to breed, just to be on the safe side.

Signs that your goat is in heat are a wagging tail, fighting, trying to mount another doe, or letting one mount her, clear discharge from the vagina, or bleating for no apparent reason.

Gestation for a goat is five months—approximately 150 days, give or take one or two.

Goats can give birth to five kids in any one litter, although the average is two to three.

Many people allow their goats to breed once a year to stop the milk supply from drying up.

You can milk a pregnant doe, but it is best to let her dry up around two months before the birth. This will allow her body to rest and build up nutritional reserves for her kids.

A doe will breed for as long as she is alive, generally around 10 to 12 years, but the older they are, the more chance complications will occur.

A doe can also become pregnant she is lactating.

Breeding Season

The fall is breeding season. All does want to breed and will have no qualms about communicating this desire to you but you need to remember that only your best specimens should be bred no matter how much they try to tell you otherwise. Goats can breed as early as two months, but the typical time is around four months. Most goats will experience their heat cycle in the fall. The safest time to breed is after the seven-month mark, with an approximate kidding time of around one year.

A few maintenance-related items need to be taken care of before breeding season. You will want to make sure they have had a Bose shot if your region is known to be selenium-deficient, trim their hooves, and conduct both a fecal analysis and a CAEV test. You will also want to clip at least their stomach region. Not only will this make your breeding season easier, but it will also allow you to give your goats one more once-over before they breed to make sure they are healthy and not demonstrating any symptoms indicative of disease. If they feel a bit thin, you can also take this time to address any deficiencies in their diet.

Most goats breed seasonally, but several miniature breeds can reproduce throughout the year. Even when this is the case, heat is still most apparent in the fall. As the daylight decreases when summer turns into fall, it will begin to go into heat on a three-week cycle. If your doe has continuously short cycles, you need to contact your vet to see if there is an issue. As the daylight decreases, the males begin to get into a rut. These changes will cause your goats to become a bit restless. As you see this happening, make a breeding plan to be well prepared when the time comes.

A doe in heat will have a red and swollen vulva with vaginal discharge. They will also wave their tail rapidly. Does in heat will also be very vocal and act more like a buck than they usually would. Also, expect your doe to produce less milk during this time.

The odor associated with bucks in rut originates from the buck urinating onto his face and into his mouth. The reason they do this is to attract the does and allow the does to find them. They will also make very distinct faces and sounds. The fighting amongst other bucks increases in frequency during this time . Mounting also becomes more frequent. Bucks become concerned with nothing other than mating, which includes a lack of concern for eating. Make sure to supplement their diet to make sure they maintain their health.

Once a doe has shown signs of being in heat, you can place her with a buck. They will dance around each other, urinating and making sounds a little at first. Once they mate, it lasts only for a few seconds. They should do this a few times to make sure that the doe is impregnated. You then watch to see if she goes into heat. If not, you will know it succeeded. Sometimes, a doe simply refuses to mate with a specific buck. This can happen when an adult doe is put with a young buck. Once kids are born, the buck will need to be separated from the doe and the kid. For one thing, his presence increases the risk she will be re-bred far too soon.

Because bucks need to be separated during certain times and can be a pain during other times, many goat owners simply choose not to have one. Instead, they will pay for buck service to get their doe's bred. There is paperwork required for this arrangement, but it is fairly common.

Knowing whether your does are in heat is a bit more challenging when there are no bucks present. One solution is to use a buck rag. This cloth has been rubbed onto a buck in a rut and then stored.

If the doe is in heat, she should become excited and make noise and rubbing against the rag. This will let you know when the best time is to have a buck brought in. Alternatively, you can bring her to the buck or have her artificially inseminated. If you are having a buck brought in, one method is to lease the buck. If you plan to do things this way, you will need a separate enclosure for the buck away from your does. The buck then stays with you and your doe for as long as you feel necessary. The cost of leasing a buck should be much lower than that of keeping one all the time. The minimum amount of time the buck should spend with you is around three weeks. This allows them to experience an entire breeding cycle. If this does not suit you, you can often find buck owners who will allow your doe to come and stay with them. For those who do not have the luxury of time, there is always driveway breeding. This involves a prior arrangement with a buck owner to be available once your doe comes into heat. Then, when she shows signs, you bring her to the farm where the buck is. Upon arriving, you exit your vehicle with the doe on a leash, and the process happens without even removing the leash so you can quickly load the doe back into the car and head back home.

Artificial insemination requires the collection of semen from the buck and proceeding to put it into the doe's reproductive system. This is convenient because the semen can be frozen and ready for the doe whenever needed rather than coordinating schedules, but this is not typical practice due to the cost. Still, high-quality breeding will occur this way. The cost can be managed by going in with other breeders and sharing the equipment. There are also semen collectors who can pass through and can retrieve the semen for you.

Immediately after breeding, there is not much to do. It does do not even show much during the first three months of the pregnancy, and milking can continue during this time. Her feed can stay the same initially as well. After three months, it will need to be adjusted. Goats gestate for around 150 days. This being said, it can vary slightly, so it is a good idea to mark your calendar 145 days out just in case. Know false pregnancy, a somewhat common occurrence in goats. This is when the goat exhibits all the signs of being pregnant, but when it comes time to give birth, fluid is released, but no kid is birthed. This is frustrating but not necessarily dangerous.

The majority of pregnancies only require basic care, like good nutrition and clean shelter, but there are still instances in which complications overrun a pregnancy. This can cause an abortion where the pregnancy cannot continue to completion. If the pregnancy continues until term, but the kid is dead at birth, it is called a stillbirth. These two things can happen for a variety of reasons.

If a kid will be born with genetic defects or malformation, they are typically aborted in the early stages of the pregnancy. They often occur so early on that you may not have even realized the goat was pregnant. There is nothing you can do to prevent this from happening. An overly stressed goat will also experience problems having a healthy pregnancy. Stress can occur because of anything from the weather to a lack of nutrition. You have more control over this cause of abortion and should work to eliminate anything stressful by making sure your pregnant goat is protected by a clean shelter and well-fed with a balanced diet. You should avoid moving your doe toward the end of her pregnancy, which is likely to cause her significant stress. Half of all goat abortions are thought to be due to infections. Not only can an infection affect one pregnancy, but it can also spread throughout the entire herd and affect other pregnant does. If multiple abortions occur, you will need to take the fetus to the vet to be tested with a necropsy process. An unsuccessful pregnancy may also be due to poison or injury. If a goat consumes the wrong plant or medication during pregnancy or is butted in the wrong place, it may abort.

Hypocalcemia occurs when a goat experiences a calcium deficiency from not receiving enough calcium in its diet to support itself and the kids it is carrying. This happens to a doe near a pregnancy's culmination or even during the lactation phase. A heavier producer of milk is more likely to experience this condition. Goats experiencing hypocalcemia will no longer wish to eat anything, especially grain. The doe becomes weak by not taking in enough food and may run a fever or become depressed. Here, weakness is caused by a lack of calcium in the muscles. This can be avoided by ensuring the doe is consuming a proper diet throughout the pregnancy and lactation. Offer alfalfa to help boost calcium. You may need to watch how much grain is given to make sure she takes in enough alfalfa. The ratio should be around two to one in favor of alfalfa. You can provide alfalfa in the form of pellets, so they can be fed alongside grain.

If you discover your goat is experiencing hypocalcemia, you will need to administer Nutridench as soon as possible. This will provide the energy needed to continue the pregnancy or continue providing milk. After this, you will want to contact your vet to develop a more long-term recovery plan. This usually includes a prescription with potassium, calcium, phosphorus, and magnesium. If the doe is dehydrated, it may also require intravenous fluids. Once the goat's heart rate returns to its usual state, you will know it is on the road to recovery. You will probably need to continue administering the medication throughout the duration of the pregnancy. Does who are within a week of kidding and do not quickly recover can be given a prescription called Lutalyse, which will induce birth.

Hypocalcemia is often accompanied by a condition called ketosis, in which a goat is not getting enough energy. This happens when a goat stops eating, which causes a metabolic imbalance causing the body to release fatty acids. The liver typically employs these fatty acids, which produce by-products known as the ketone bodies. Overweight goats experience this condition more often and may experience it during the early stages of the pregnancy. A key identifier for goats experiencing ketosis is a sweet odor from their breath, compounded with hypocalcemia's usual symptoms. Ketosis should be treated with Nutridench.

The final few weeks of a doe's pregnancy will require added care and maintenance to guarantee a smooth birthing. To reduce the risk of abnormal labor and white muscle disease, anyone living in a region known for selenium-deficiency should administer a BoSe shot. Those who vaccinate their goats should administer a CDT shot. This will allow the kids to build up immunity from both tetanus and enterotoxaemia. The tail area will need to be trimmed and the area around the udders. This helps the doe remain clean and makes it easier for feeding. You should not milk your doe during the final two months of her pregnancy. Make sure to be subtle about any changes to diet during this time.

A kidding pen will need to be set up and ready a few days before day 145 of the pregnancy. The pen should be sanitized using bleach and water. A new layer of wood shavings or straw should be put down as bedding. It is also helpful to have a clean, sanitized bucket nearby. You may also choose to place a baby monitor near the kidding pen so you can be alerted of the birthing process.

Preparation will make the birthing process much less stressful. To make sure you have everything you need and know where to find it, you may want to put together a kidding kit. The kit should have 7% iodine, a prescription bottle, flashlight, floss, bulb suction, towels, surgical scissors, gloves, and a syringe. It is also helpful to have a list of important phone numbers for people like your vet or other goat friends who can help. Betadine surgical scrub should be procured to help you while washing the goats, and an obstetrical lube is always a good idea. Do not forget to have an empty bottle with the right teat if bottle-feeding becomes necessary. When the time comes, you will need a bucket of hot water or a hot water source near the kidding pen. If the doe needs assistance, dish soap can be used for your hands and the doe's vulva to clean off the lubrication.

Usually, the doe can birth with no help from you, but you will still want to be around if an emergency occurs and to make sure the birthing process goes as smoothly as possible. When your doe is nearing the time for birth, its tailbone changes shape as it rises, and the ligaments connecting the pelvis stretch. If a hollow area forms on the sides of the tail, this could indicate that birth is near. The most accurate measure of how soon the birth is likely to occur comes from feeling the ligaments on the side of the tail. Typically, these ligaments are firm, but they will become soft and will not be perceivable. Once you can no longer feel them, you can be fairly certain the birthing will happen within the next 24 hours. When they become soft, move her to the kidding pen. When a doe is getting ready to kid, you will also experience a few behavioral changes. The doe will isolate itself, have vulva discharge, lose its appetite, become aggressive or restless, and its udder will become firm and shiny. Once the process has begun, you can leave the baby monitor on and allow her to focus on kidding.

Basic Kidding

If you are a new goat owner or have had your goats for a year or two and just bred them for the first time, you may be worrying about them getting through kidding safely. The main thing to remember is that having a baby is normal and, most of the time, it will go just like nature planned.

Goats usually deliver their kids between 145 and 154 days. Use 150 days to estimate kidding, but keep a close eye on your doe starting at about 144 days.

According to David MacKenzie, from *Goat Husbandry*, as long as you can see the kid(s) as a bulge on the right side and see movement, the goat is unlikely to kid within the next 12 hours. Kidding, or parturition, is divided into three stages:

• The first stage of labor is when the uterine contractions dilate the cervix by forcing the placenta, fetus, and amniotic fluid against it. This can last up to 12 hours in first-time moms but is often faster for those who have previously kidded. Again, every doe is different.

• The second stage of labor is when the doe pushes the kid(s) out. It usually lasts less than two hours but can be longer.

• The third stage of labor is the expulsion of the placenta and the reduction of the uterus back to its normal size. Usually, the placenta is passed within an hour or two after birth, but it can take hours in rare cases. The uterus does not reach its pre-pregnancy size until about four weeks later.

The first stage starts with estrogen secretion by the ovaries, which causes the uterus to contract.

You will not feel the kids moving. The bulge in the doe's right side will change, and the rump will slope more. This may not be visible to any but the trained eye.

You will see restlessness begin in the doe. If you have a clean kidding pen prepared, now is the time to move her there. Like all mammals, goats like a quiet, safe place to have their kids. It should be lit well enough (or have access to light) so you can see what you are doing if you need to help but dim enough to be comfortable. The area shouldn't be too small, so she can move around as the labor progresses.

Avoid putting water in the pen, as kids have been known to drown in it. To give the mother water, make sure the water is warm and once she is finished drinking, remove the water from the pen.

Around this time, you may see a thick discharge. This means that the doe has lost her cervical plug. You will likely see a change in discharge as labor progresses. It thickens and changes color and can be blood-tinged; this is normal.

What is not normal is thick, rusty-brown discharge, which may indicate a dead fetus. If you have questions, contact your veterinarian or an experienced goat breeder.

Your doe, at this point, will probably reposition herself regularly, trying to get comfortable. She may lick herself or objects, "mama-talking" (a special talk reserved for welcoming kids), or with a very spoiled goat, demand you stay there and pet her throughout.

The second stage of labor is where the real work begins. The babies have lined up for birth, and the doe pushes them out, in sync with the uterine contractions. The contractions become stronger and closer together.

Some goats deliver standing up, while others prefer lying down. The doe might cry out at this point. It depends on how stoic she is. The first sign that tells you the labor is progressing is what looks like a balloon at the vaginal opening. This is the membrane surrounding the baby.

The doe may start licking in earnest between pushes, sometimes situating her body so she can lick up the amniotic fluid. With more pushes, you may see two little hooves and a little nose, which indicates that the baby is positioned properly. The kid is moving down the birth canal.

If you see just the nose and no legs, and the birth's progress seems to have stopped, insert a thoroughly washed finger in to feel for bent back legs. You sometimes need to pull just one of these gently up to help the baby get out; with others, it may take two. If you pull one leg slightly forward, it will decrease the shoulders' width, and the kid should come out easily now with just another push or two.

Anytime you have to help a goat and put your hand in her vagina, it is important to have clean hands and short nails. Ideally, also wear gloves. Often goats, especially minis, are born in a breech position - back feet first - with no problems. Frank breech position, where the hind legs are folded underneath the kid, is potentially a bigger problem, but small kids can also be born this way. Otherwise, it will need to be corrected before birth, which you can do by gently pulling the feet and then the kid out. This prevents it from accidentally inhaling amniotic fluid and getting aspiration pneumonia or drowning.

Another presentation problem I have encountered only once out of hundreds of births is crown presentation. This is where the kid's nose is pointing down toward the body, with the top of the head presenting. Because I didn't know what I was feeling and the vet's hands were too big, we had to perform a c-section. (That kid was born four hours after his brother and did just fine.)

Another unusual position is transverse, where the kid is sideways. This will always stop the birth, and the kid has to be turned with back legs coming first and gently pulled out. Once a kid is born, wait for the umbilical cord, if it hasn't already broken. Once the cord breaks on its own or collapses when the blood flow stops, you can tie it off securely with dental floss in two places: an inch or two from the kid's belly and an inch past that. Only now should you cut it.

During this time, the mom will be licking and cleaning the baby. If the doe does not want to get up or can't reach the kid, you can fetch it for her. She will continue with this behavior until the next kid is ready to be born, which can be quickly or can take another hour. Longer times may be a sign of malposition, so if a placenta has not been delivered yet, and you aren't sure if there are other kids, you may want to check. Remember to err on the side of not intervening unless needed. This is where experience comes in.

There are a couple of ways to check for more kids: First, you can check inside the doe with a finger. That will at least tell you whether another kid is in the birth canal and needs help with positioning. If that tells you nothing, you can "bump" the doe. Stand behind her, and with your hands on the doe's abdomen, lift up quickly to feel for another kid. An effective but more invasive method is to check inside the uterus with a well-washed, lubricated hand and forearm. I have found that having a bucket of soapy water helps this effort immensely. Wash the perineum and be gentle with your exploration. A loose-feeling uterus will contain no other babies.

I have had to do this only once in my seven years of kidding experience. There, the doe had a ring womb, which means the cervix will not dilate enough, and I had to slip the cervical lip around the large kid's head. Usually, you will know that doe is through kidding.

If you deliver a kid that is not breathing and seems very weak, you can try baby CPR or simply hold tight to the kid (one hand on the leg and one on the neck to stabilize the head) and swing it back and forth in a 90-degree arc to clear the mucus. This is what I did with the "c-section" kid, born four hours after his brother. If the kid cannot suckle, you may need to tube feed it.

Once the kids are born, the mother should nurse, which causes a release of oxytocin—also known as the bonding hormone. It not only helps mother and baby bond, but it stimulates uterine contractions that lead to the delivery of the placenta and closing of the cervix. You will sometimes need to help the kids find the teats to nurse; in rare cases (once in my experience), the mothers will not know to nurse their young. Breeders who pull the kids at birth should milk the goat, as this has the same effect. There is normally only one placenta for each litter, and it comes out after the birth. I understand that more than one placenta may exist with some goats, and it may be expelled between deliveries of kids. Expect to see that the doe has a bag of amniotic fluid attached to the umbilical cord hanging from her vagina. The weight of the fluid helps to pull out the placenta after it detaches from the uterine wall.

Failure to deliver the placenta may indicate that another kid is still inside the doe. Never pull on the membranes to remove the placenta as it can cause ripping and lead to problems later. The placenta is not considered retained in a goat until at least 24 hours have gone by. You may obtain a prescription from a veterinarian for oxytocin for a retained placenta, but do NOT routinely use it. Do not assume that if you found already-born kids and did not find a placenta, it is retained. Goats, like all mammals other than humans, typically eat the placenta.

Once kids are born, dip their navels in 7% iodine to prevent navel problems. Make sure they are thoroughly dried, especially if the weather is inclement. They need to receive colostrum within the first hour, if at all possible. Once mom has completed her job, I have a ritual of bringing hot oatmeal with molasses and a bucket of hot water to offer her. The water replenishes her system, and the oatmeal is a great treat with the added benefit of being galactosemic (helping to produce milk).

Chapter 9: Seasonal Herd Care and Maintenance

In the world of livestock, goats are the easiest to care for. Given the right conditions and the right feed, your healthy goats may seem self-sufficient. Although they may be easy to raise, they still require care. In this chapter, you will learn what care is required for raising and maintaining healthy and happy goats, from trimming their hooves to castration.

Grooming

We brush our teeth and hair, take baths and eat right, so why should our goats be any different? There are two different types of grooming. One is grooming for health and wellness. Routine grooming will help you develop a bond with your goats and allow time for health inspections. The second type of grooming is for shows. Grooming for shows is a lot more involved—regular brushing, bathing, and shaving—since they need to look their absolute best. This section focuses primarily on everyday grooming.

Petting Your Goat

Your goats will beg you to give them a good scratch, just like Fido does. Just remember, goats don't like being petted on the head. They get spooked when you try to pet them on the head because they can't see what you're doing. Try petting your goat on the back, chest, or neck. Besides avoiding the top of their head, try not to push against their forehead. While it's funny and cute when they are little, they are training to knock you off your feet when they get big enough to establish dominance.

Brushing and Bathing

Goats love to get a good back scratch. You will see them brushing up against a tree trunk, a wire fence, or the side of a barn to scratch that itch. We like to take the heads off stiff-bristle brooms and screw them to trees so the goats can brush up against them, and they love it.

When you spend time brushing your goats, you create a grooming bond, similar to the one a mother makes while grooming her young. You can also take this time to inspect your goat for injuries, abnormalities, and hooves in need of trimming.

During the colder months, goats grow a winter coat. You will see what looks like dryer lint close to their skin. This is their winter fluff. Don't try to brush this out when it's still cold, as they need it to keep them warm, but you can help groom them in the spring when they start to shed their winter coat.

To brush your goats, you need a stiff brush and a soft brush. The stiff brush helps get all the old winter fluff off, while the soft brush is used for daily or weekly brushing. Brush their fur in the direction in which it is growing, and try to brush their chest, back, and legs.

If you are raising goats for livestock only, there is little need for bathing unless a goat is sick or gets stuck in the mud, or you just want a better-smelling goat. I recommend reserving the task only for times when it is absolutely necessary. Too much bathing can interfere with the natural oils that keep their skin and coat healthy.

Remember, goats hate water. Bathing your goat is not a fun task for either you or your goat, but they are creatures of habit, and if you bathe early, they will get used to it.

If you do want to bathe your goat, I suggest using a collar with a short leash and tethering them to a fence or a milk stand (stanchion) for this task. This holds them still while you remove dirt and shampoo them.

Bucks smell worse than does do due to the buck "cologne" they create during mating season. This is a sticky, smelly residue that is quite hard to get off. You want the cologne to remain during the mating season since the ladies go wild for it. The buck smell does dissipate after the mating season has ended.

Use goat milk soap or livestock shampoo sold at farm supply stores. Wash your goat on a warm day when they will have plenty of time to dry in the sun before nightfall. Avoid cold, wet, or windy days for bathing.

Hoof Care

As daunting as it may seem, I recommend that all goat owners learn to trim their herd's hooves. You can hire this job out to trained people, and that may be the best choice for you, but this is a task I am confident you can learn to do on your own with practice.

When a goat lives in its natural environment, rocks, forage, tree bark, and the like keep its hooves trimmed. When you place a goat in a fenced-in pasture, performing routine tasks to care for their hooves is essential in protecting their health. Overgrown hooves can lead to leg, joint, muscle problems, and footrot, which happens when bacteria is trapped in the fold on the hoof.

Think of a goat's hooves as you would think about your fingernails. The growth past the skin is what you need to keep trimmed and clean. Goats have cloven hooves and one dewclaw (on the back of their ankle). How often you must trim your goats' hooves depends on the individual goats and their living conditions. It is best to check your goats' hooves weekly to help determine how fast they grow. A general rule of thumb is to trim them every two to four weeks.

When you buy your goats, ask the breeder to show you how to trim hooves on one of their goats. Take pictures or notes about the process so you can refer to them as needed. If your breeder cannot demonstrate, you can contact your goat vet to help you through the process until you are comfortable trimming them yourself. Goats are very skittish and jumpy. The last thing you want to do is cut your goat or yourself. Be calm, talk to your goat through the whole process, give it time and be patient. Do a few practice rounds before performing the real deal by walking your goat to the stanchion, securing them, lifting each hoof and trimming it, and then returning them to their usual pasture.

How to Trim Goat Hooves

Skill Level: Beginner | Estimated Material Cost: $10 or less | Time: 10 minutes per goat

Supplies, Tools, and Steps

- 1 cup of warm water
- Small stiff-bristle brush
- Stanchion or a collar and leash
- Feed or treats, as needed
- Stool or bench
- Clippers or hoof trimmers
- Antiseptic spray
- Styptic spray or powder

1. If the goat's hooves are muddy, soak them in warm water and use the brush to remove any mud.

2. Secure the goat to the stanchion. Offer the feed or treats to keep them occupied during trimming. Place the stool beside the goat (not behind), working on just one hoof at a time.

3. Using warm water and a brush, remove any dirt, and clean the hooves.

4. Spray the clippers with antiseptic spray. Do this between each trimming to prevent spreading infection or disease from hoof to hoof.

5. Grab the hoof and bend it back toward you; do not raise the leg forward to trim.

6. Using your clippers, trim any excess growth away from the pad of the hoof. The pad is soft and pliable, and the outer edge is the part you will be trimming. Make sure the hoof is level and straight. Any curves should be trimmed.

7. If you accidentally cut the goat, use the antiseptic spray and the styptic spray or powder to stop the bleeding.

8. Dewclaws need not be trimmed often, usually only on older goats or when the dewclaws start to curl into the goat's skin.

Troubleshooting

Not cutting enough or cutting too much can affect the way the goat walks and can cause issues. It's important to make sure their hooves are trimmed nice and level.

To Dehorn or Not to Dehorn

Dehorning is the act of removing a goat's horns permanently. Dehorning goats is the subject of much debate among goat owners, and both sides have compelling arguments. I have been, and I still am, on both sides. I have goats with horns, goats that have been dehorned, and naturally polled goats, and I've even owned a goat with scurs — we'll get to that in a minute. Let's start with the basic pros and cons of horns.

Horn Pros

Horns help goats defend themselves against predators and protect the herd. This ability is diminished in dehorned goats.

Horns help goats establish a natural pecking order. A goat with horns will dominate a goat without horns.

Horns help regulate body temperature. Certain goats, such as Angora goats, should never be dehorned because, without their horns, they could overheat and die.

You avoid the risks associated with dehorning. If not done properly, dehorning can cause permanent brain damage or infection.

Horn Cons

Horns can get stuck in fences and cause injury. This risk is diminished in dehorned goats.

Horns can injure you and others. Dehorned goats are less dangerous to humans.

Goats with horns are often harder to sell than dehorned goats.

Goats with horns rarely are allowed in shows. If you plan on showing goats in the future, your goat needs to be dehorned.

Dehorning, or disbudding, is usually done when the goat is just a couple of weeks old. This is a medical procedure during which a hot iron is used to burn the horn buds (horn buds are the beginning of the horns forming) off their head. If you want a goat without horns, let your breeder know before purchasing your goat or contact your vet to perform the procedure.

Polled goats are naturally hornless. A polled goat comes from a parent or parents that are polled. No disbudding or dehorning will ever be needed if you have a polled goat.

Sometimes dehorned goats develop scurs or partial horns that grow after a goat has been disbudded or dehorned. This can happen months or even years later. Generally, these scurs break off naturally, but sometimes they will continue to grow, which is fine. I don't recommend removing the scurs during older age unless it is causing a health issue, and if that's the case, it's time to contact a vet.

Chapter 10: Selling Meat, Dairy, and Other Goat Products

Running Your Side Business

If you plan to sell goat meat, it's important to market your meat to reach your buyers. You have likely seen products with labels like "organic," "all-natural," and "grass-fed." As these labels become more popular, so do the licensing requirements and regulations to use them. You can no longer say something is organic without being a licensed organic farm, which requires fees. If your farm is organic, but you don't pay the fee, you can't call your products organic.

There are still honest marketing approaches you can use without paying to reach your target customer. For instance, my farm friend who lives out West, where there are many devastating fires, advertises, "These goats were raised to maintain a fire-safe environment," which appeals to her local market.

There are lots of ways you can market your product to find your own niche and draw in the crowds. Here are selling points you can think about for marketing your goats:

- Family-farm raised
- Pasture-raised
- Free-range
- Humanely-processed
- All-natural
- Locally-produced
- Farm to table

Although goat meat is widely consumed in other parts of the world over beef or poultry, it is fourth on the list in the United States, except for ethnic and specialty markets. Certain ethnic groups—such as those with heritage from Asia, the Middle East, Latin America, Africa, and the Caribbean—will probably be your largest consumers.

Religious celebrations where goat is on the menu for celebratory or ceremonial meals will often encourage a spike in demand. Thankfully, there are many celebrations throughout the year, giving your business a year-round market. A few of these holidays include:

- Chinese New Year
- Greek Orthodox Easter
- Rosh Hashanah
- Islamic New Year
- Start of Ramadan
- Passover

To find your customer base, you need to go where they are. Start by advertising at specialty stores, farmers markets, local restaurants, religious centers, and even livestock auctions. Create a desire in people you already know by inviting them over for dinner and letting them see how good the goat tastes. Once people see and experience the difference in farm-raised meat treated ethically and humanely, you can create your own market. Help educate those in your community about goat meat's benefits, how lean it is, how sustainable the livestock is, and the importance of cutting out the middleman and getting their meat direct from the source.

Remember that on-the-farm slaughtering for sale is illegal, and you can only sell the live animal—what we call "on the hoof." The purchaser can then transport the animal to the slaughterhouse or have a mobile butcher process it.

Chapter 11: 7 Deadly Mistakes New Goat Owners Make (and How to Avoid Them)

Over the past few years, I have encountered people who had made a mess of enormous capital invested in goat farming. Investing so much is not the problem, but the problem is when you are beginning to lose everything you have invested without the desired output over the years. My parents made mistakes while raising animals; it was as if everything was going against us, and at that time, we thought it was a spiritual attack to jeopardize our efforts. When you are not well equipped with updated knowledge of raising an animal, you will likely lose the animal and lose your investment.

Is making a fortune your driving force toward goat farming? This marks the beginning of failure in the business. Instead, focus on quality production, and you will make good profits.

The business's failure rate will be reduced by at least 70% if you avoid these mistakes that first-timers make in goat rearing.

1. Getting Started with a Single Goat

Goats are peculiar creatures - they are social, curious, and intelligent. Just as most humans do not enjoy being alone for very long, goats easily get bored and lonely. Goats are not dogs or other animals you can rear singly. It is never a good idea that you begin a goat business with one goat. At least two goats are required to get started. Apart from the mating aspect, goats love the company of other goats around them.

You can begin with a doe and a buck or even a doe and wether but make sure they have at least one companion.

Goats are herd animals. They depend upon staying together for safety. They have few natural defense mechanisms but many predators.

A lonely goat will try to escape or climb and get into your garden, which can lead to devastating effects.

Note that a single goat will be a noisy goat, as it will always call for a companion.

If you purchase a goat, be prepared to add a companion goat.

2. Combining a Buck with a Milking Goat

Bucks are generally stinky and have a characteristic odor for about half of the year. Remember this before buying a buck; the buck's smell will undoubtedly get into the milk. Smelly milk will undoubtedly influence the market value of your milk and its derivative products.

I have received several complaints from goat farmers about the milk quality, while some complain that the milk has a foul taste. The first question I ask is if they have combined a buck with a milking goat. Most of the time, I get a YES response. The solution is straightforward as it lies in separating the buck from the milking goat. If you keep a buck with a milking goat on the same farm, then it must be spacious enough to keep them well separated.

3. Poor Breeding Technique

Inadequate knowledge of breeding can lead to an extremely undesirable output. When breeding, avoid breeding a larger-framed male goat (buck) with a moderately sized female. When a female goat is not sexually matured enough, avoid breeding it against a matured buck as this can cause complications during the birth of the new kid. Again, the young doe could die during the process of parturition due to tears. I recommend first-time goat farmers consult a professional when they want to embark on the breeding process.

4. Poor Market Research

For any business anyone wants to delve into, intensive market research is a key to profitability and sustainability. Before purchasing a goat, it is compulsory to determine the demand for goats in your locality. If you are rearing a goat to make money, you must see goat farming as a business and treat it.

Whenever I receive complaints of poor sales of goats and their products, I ask if thorough market research was done before embarking on the business. If the market demand in your locality is very high for meat, you shouldn't focus on raising milk-producing goats. More importantly, you shouldn't seek advice from other breeder producers different from yours, as applying their techniques to your herd will cause health problems for your goats.

Also, you need to check your zoning regulations and whether it permits goat rearing where you are not living on a farm out in the country. You may not be allowed to rear goats. This is one reason why proper research is needed before delving into the business.

5. Inappropriate Breed for an Environment

Goats are primarily dry climate animals, although there are breeds that show more resilience and seem more adaptable to varying climatic conditions than others. It is imperative to find a breed of goat that fits your climate and other environmental factors.

A goat that does well in a particular region does not mean it will do well in all areas.

6. A Jack-of-All-Trades in Goat Business

I tell first-timers going into the goat business that a goat business is not as simple and easy as many think. It is crucial for someone new to goat farming to study the various breeders according to the economic and market demand. To start a goat business on a small scale, do not try to produce breeding stock, show goats, or slaughter goats all at once, as you will get frustrated eventually.

When a client approaches me to help them set-up a goat farm, I evaluate many factors and then come up with a recommendation. One recommendation is to start with one aspect of goat breeding and become well acquainted with the particular breed's modus operandi. When you become familiar with a breeder, you can easily maximize their potential to your advantage.

7. An Urban Approach to Raising a Goat

If you are considering rearing goats with an urban approach, then do not try it. I have had many years in the integrated and sustainable agricultural system, and I can tell you from experience that animal farming is much more complicated than crop farming; therefore, you need to understand how to do things and not rely on your intuition as it likely will fail.

I have seen cases where novices in the goat farming business attempt to confine their goats to a particular spot to avoid them moving around. In other cases, some restrict their goats to a small building. Goats were not created to stay indoors, so give up on your urbanite approach to goat farming.

I need to reiterate that goats are typical livestock. They are not cats or dogs, so they are not meant to live in the house with you - they are created to live outside. The consequence of confining goats is that they become unhealthy and die due to worm infestation and disease.

Important Notes for Goat Farmers

- Goats don't like getting wet, and because of this, they do not thrive well in moist, swampy areas. They require a dry shelter and dry paddocks.

- Dairy goats get upset when you frequently change their routine. They do not like it when you rearrange the milking stands too often.

- Goats are typical browsers; they prefer hay, bushes, and trees to grass. Do not expect them to mow your lawn.

- Aside from goats being clean eaters, they eat a lot. They do not eat contaminated food, and they do investigate what they eat. Prepare a sufficient budget to meet their feeding demands before importing them to your farm.

- Goats of different breeds have distinct personalities and traits, so do not expect all breeds of goats to behave the same way. Always do proper research before purchasing a goat, so you know their distinct behavior and determine which temperance suits your preference and personality.

- Goats are good listeners. If they trust you and are inclined toward you, they will respond to your call and can also call out to you whenever you appear.

- Pasturing a buck with a milking goat is a wrong move. The smell of a buck can make the milk taste bad.

- If you keep a milking goat, milk it at least once in 24 hours. You cannot afford to leave a milking goat and go on vacation.

- Check local zoning regulations to know if you can have goats out of a farmyard in your country.

- Before you embark on the goat farming business, take your time to locate sources of medical help closest to you because this will save you time when you need a prompt response from a goat vet.

- From experience, any species that experience sexual maturity too early, short gestation, and multiple births are likely to die earlier than envisaged, no matter your intervention.

- Call a vet doctor during emergencies.

Chapter 12: Tracking Your Business's Progress

Several factors control the beneficial activity of goat farming. A privileged position to enter goat farming is a specialized showcase, unlike the most cited and used farm animals such as beavers, pigs, and chickens, which means there are less committed competitors. Raising goats also requires less funding than other four-legged farm animals and will give a higher profit. Another thing is that you can create multiple products to choose from. Is it true that you are interested in selling dairy products like milk and cheese? How about selling your hides to the calfskin industry? Goat meat is also a decent decision, especially for the increasingly extravagant market, which requires goat meat as a must. The sale of goats can produce big profits. Regardless of what market you hope to be in with that goat business, here are tips for a productive goat farm:

Breed—First, know which exact breed of goats is best suited for the goat-breeding barn you want to have. Even though they are all goats, they are not all are the same. Some goats are even bred for certain purposes, such as meat creation, while others are better at producing dairy products. So, it is better to have a Boer goat if you are concentrating on milk and not on meat. After the basic purchase, go for acceptable quality goats. Choosing the right breed is the pathway to your business success.

Appropriate consideration—Learn to handle your goats properly and raise them well. A goat shelter and pen are a necessity, so look at it as an investment and insurance. Also, feed your goats with common food or food proper for their purpose and job. For example, use feed specifically indicated for a kid or lactating goats. The benefit of using feed is that the nutritional requirements are already taken care of, and you don't have to worry about overfeeding, but there are more expenses and even hardware needed for feeding. Going for little by little characteristic foods is an increasingly beneficial goat breeding technique, but you still need to know that what you're feeding your goats is safe for them and still healthy. What's more, never neglect to use a veterinarian's services to make sure that your animals are in the best shape and protect them from any disease.

It is very important to have business plans for goats if you decide to invest in this market. The investment required to raise goats is much less compared to other larger animals, such as sheep and bulls, and the yields that can be produced from this are realistically acceptable. Running a goat farm is not a walk in the park, but it is achievable, especially for those who are smart enough to prepare and know what they're doing to keep things running smoothly. For those interested in this business, we will see a step-by-step strategy for a livestock business plan:

- Before thinking about investing in goats, sheep are versatile animals, so you need to know the negative factors when raising goats. The goat is a good source of many things, such as:

o Meat: Very popular in some target markets.

o Dairy products: Milk and cheese. Usually, the more distinct something is, the more expensive it can be, thanks to supply and demand.

o Fiber: Goatskin is truly an incredible source of textiles, such as cashmere.

o Goats: Why kill them if you could sell them? Goat farming is an industry in itself. Goats are easy to breed, and a single goat can have a high value on the market, depending on their condition and breed.

• After choosing the type of goat for sale, it is time to move on to the next step in your goat breeding business. Choose a goat breed that matches your needs because not all goat breeds are the same. For example, if you like meat, choose Boer goats because they are bred specifically for that purpose. If you want them to produce fleece, look for cashmere goats, but if you are in the early stage of crossbreeding, go for Kiko goats. The list goes on. Choose carefully because this is an important factor in the success of your goat business.

• The next step is an urgent step to implement business plans for raising goats.

Before you begin, think about your current financial plan or how much you can afford to invest. Besides the initial capital, raising goats requires a large field, goat barn, and goat food.

The establishment of a goat farm favors the continuous development of livestock activity. When you start a goat farm, learn about the types of goats you have.

Evaluate the nature of the goat that fits the bill for the delivery of the meat. Gather those who are suitable for this field. Some goats can produce fibers. For the goats used to create milk, they must be set up in the farm area with maintained drainage tools. Goats to be used for meat delivery must be properly cared for. The goat slaughter and slaughter program must be respected to allow the creation of immaculate and high-quality meat. Improper feeding, aging, and slaughter of the creatures will lead to the taste of the cooked meat being unpleasant.

Choosing goat farming may feel extreme at first. Yet, with the best possible information, commitment, and hard work, a goat farming business can be a wonderful thing.

Conclusion

Raising goats is not for everyone. Let's face it, what sounds like a great idea, in theory, can often end up becoming overwhelming in reality. If you don't have the time, or the ability to commit yourself, or the discipline, then goats might not be for you. There is no shame in recognizing your limitations, and only nobility in avoiding actions can eliminate trouble for yourself and other living beings.

But if you have the time to commit, have space and money to spend on the fencing and feed, and the time to build a shelter, raising goats can be very rewarding. If you love animals, you know how gratifying it is to see your animal happy and contented, thriving and enjoying your interaction. Goats love company and are very social creatures. They are generally glad to see you when you enter their enclosure and rarely resort to the head butting you see in the cartoons.

As more people leave cities and move to rural areas, consciously wanting to reconnect to the land and the old ways of doing things, more people are raising farm animals like goats, chickens, and even pigs. While some may find it was a great notion that didn't pan out, others may find great contentment, an inner peace that can only come from becoming connected to the land on which you live. Raising goats can be a great way to do that. In a few months, when you're pouring goat's milk onto your cereal one morning, milk that came from goats you raised, cared for, and milked, you will feel a tremendous sense of accomplishment and completeness. More and more, that sense of completeness is missing from daily life for most people, but for those who get back to the land and try something new, like raising goats for milk or fiber, it is a different story. The path to true happiness does not always lie in more advanced technology. Talk to people who've raised goats and see what they say about it. You may be surprised.

Here's another book by Dion Rosser
that you might like

References

8 Tips to Prepare for Goat Breeding Season. (2019, July 22). Hobby Farms. https://www.hobbyfarms.com/goat-breeding-tips-prepare/

12 Popular Goat Breeds. (2015, August 5). Successful Farming. https://www.agriculture.com/family/living-the-country-life/12-popular-goat-breeds

Building the Goat Barn | GottaGoat. (n.d.). http://www.gottagoat.com/gottagoat-goats/building-the-goat-barn/

Cleanliness Is Next To Goatliness. (2015, April 21). Modern Farmer. https://modernfarmer.com/2015/04/cleanliness-is-next-to-goatliness/

DIY: Make a Free Goat House from PALLETS. (2017, August 5). Weed 'em & Reap. https://www.weedemandreap.com/make-free-goat-house-pallets/

Goat Care and Maintenance of Healthy Goats. (2020, January 23). Timber Creek Farm. https://timbercreekfarmer.com/goat-care-and-maintainance/

GOAT FARMING AS A BUSINESS: a farmer's manual to successful goat production and marketing For the Department of Livestock Production and Development Supported by: SNV - Netherlands Development Organization. (n.d.).

https://snv.org/cms/sites/default/files/explore/download/goat_farming_as_a_business_-_a_farmers_manual.pdf

Goat Farming (how to start & make money) by Business Tips Zambia. (2017, August 22). Zambia Farmers Hub, Zambia Farmers hub. https://zambiafarmershub.wordpress.com/2017/08/22/%E2%80%8Bgoat-farming-how-to-start-make-money-by-business-tips-zambia/

Goat Reproduction Preparing for the Breeding Season – Goats. (n.d.). Goats.Extension.org. Retrieved from https://goats.extension.org/goat-reproduction-preparing-for-the-breeding-season/

Hot Weather Tips for Goat Enthusiasts. (n.d.). Brazosfeedsupply.com. Retrieved from https://brazosfeedsupply.com/blog/5950/hot-weather-tips-for-goat-enthusiasts

Housing, Fencing, Working Facilities and Predators - Goats and Health - GOATWORLD.COM. (n.d.). Www.Goatworld.com. Retrieved from http://www.goatworld.com/articles/fencing/fencing1.shtml

How a young farmer developed a goatmeat business. (2018, January 22). Farmers Weekly. https://www.fwi.co.uk/livestock/how-a-young-farmer-developed-a-goatmeat-business

How to Build a Goat Barn ⋆ThePlywood.com. (2018, August 14). ThePlywood.com. http://theplywood.com/goat-barn

https://www.facebook.com/thespruceofficial. (2018). The Spruce - Make Your Best Home. The Spruce. https://www.thespruce.com/

Keeping Goats Warm in the Winter. (2018, October 31). The Hay Manager. https://www.thehaymanager.com/goat-and-sheep-round-bale-hay-feeders/keeping-goats-warm-in-the-winter/

MorningChores - Build Your Self-Sufficient Life. (n.d.). MorningChores. https://morningchores.com

Planning, G. F., & says, S. Y. N.-B. G. P. G. (2017, May 9). My top 3 picks for goat fencing that is secure and safe. Simple Living Country Gal. https://simplelivingcountrygal.com/goat-fencing-101-everything-you-need-to-know/

Preparing and Caring for Your Goats in Winter. (2015). Mannapro.com. https://info.mannapro.com/homestead/preparing-caring-for-goats-in-winter

Raising Goats: Keeping their barn clean - Boxwood Ave. (2018, October 9). Boxwood Ave. https://boxwoodavenue.com/raising-goats-cleaning-barn/

ROYS FARM | Modern Farming Methods. (n.d.). ROYS FARM. https://www.roysfarm.com/

Top 10 Mistakes Made by Goat Owners. (n.d.). Www.Lambertvetsupply.com. Retrieved from https://www.lambertvetsupply.com/wellpetpost-top-10-mistakes-made-by-goat-owners.html

Top Ten Mistakes Made by Goat Producers. (n.d.). Www.Tennesseemeatgoats.com. Retrieved from https://www.tennesseemeatgoats.com/articles2/toptenmistakes06.html

Want to Become a Successful Goat Farmer? Here are the Excellent Tips, Benefits of Rearing Goats & Making Maximum Profit. (n.d.). Krishijagran.com. https://krishijagran.com/animal-husbandry/want-to-become-a-successful-goat-farmer-here-are-the-excellent-tips-benefits-of-rearing-goats-making-maximum-profit/

Wolford, D. (2019a, July 18). Goat Breeding 101 - Weed'em & Reap. Weed 'em & Reap. https://www.weedemandreap.com/goat-breeding-101/

Wolford, D. (2019b, October 19). A Simple Guide to Raising & Milking Goats. Weed 'em & Reap. https://www.weedemandreap.com/raising-goats-milking-goats/

Printed in Great Britain
by Amazon

86217642R00068